Advanced Sciences and Technologies for Security Applications

Indexed by SCOPUS

The series Advanced Sciences and Technologies for Security Applications comprises interdisciplinary research covering the theory, foundations and domain-specific topics pertaining to security. Publications within the series are peer-reviewed monographs and edited works in the areas of:

- biological and chemical threat recognition and detection (e.g., biosensors, aerosols, forensics)
- crisis and disaster management
- terrorism
- cyber security and secure information systems (e.g., encryption, optical and photonic systems)
- traditional and non-traditional security
- energy, food and resource security
- economic security and securitization (including associated infrastructures)
- transnational crime
- human security and health security
- social, political and psychological aspects of security
- recognition and identification (e.g., optical imaging, biometrics, authentication and verification)
- smart surveillance systems
- applications of theoretical frameworks and methodologies (e.g., grounded theory, complexity, network sciences, modelling and simulation)

Together, the high-quality contributions to this series provide a cross-disciplinary overview of forefront research endeavours aiming to make the world a safer place.

The editors encourage prospective authors to correspond with them in advance of submitting a manuscript. Submission of manuscripts should be made to the Editor-in-Chief or one of the Editors.

Luis A. Garcia-Segura
Editor

Collaborative Insights on Technology and Legal Frameworks

Towards a Secure Future

Springer

Editor
Luis A. Garcia-Segura
Nebrija University
Madrid, Spain

ISSN 1613-5113 ISSN 2363-9466 (electronic)
Advanced Sciences and Technologies for Security Applications
ISBN 978-3-032-27713-8 ISBN 978-3-032-27714-5 (eBook)
https://doi.org/10.1007/978-3-032-27714-5

This Springer imprint is published by the registered company Springer Nature Switzerland AG
The registered company address is: Gewerbestrasse 11, 6330 Cham, Switzerland

Contents

Assessing Generative AI Safety for Early Learners: Not Just Safe, but Right—Reframing Risk Through Educational Purpose and Philosophical Coherence

Jorge A. Cerdio Herrán

Abstract This chapter explores how Generative Artificial Intelligence (GenAI) should be integrated into early childhood education, emphasizing both its transformative potential and its ethical and developmental risks. While Educative Artificial Intelligence (EdAI)—the educational application of GenAI—offers benefits such as personalized instruction, real-time feedback, and automated content generation, its use with young children raises pressing concerns. These include the spread of inaccurate content, the erosion of critical thinking, emotional dependency on AI agents, privacy violations, algorithmic bias, and the marginalization of human educators. The chapter argues that such risks cannot be addressed through technical safeguards alone. Instead, they must be evaluated through the lens of a coherent educational model—one grounded in philosophical anthropology and institutional values. EdAI should not be treated as a neutral tool but assessed according to its alignment with an institution's vision of the child, relational pedagogy, and moral formation. A proposed framework includes pre-deployment philosophical evaluations and post-deployment relational audits. Ultimately, EdAI's legitimacy in early education depends not on technological sophistication but on its fidelity to education's central aim: the integral development of the human person.

1 Introduction

In the twenty-first century, technological advancements increasingly not only support human learning but also begin to shape its foundational conditions. Among these, Generative Artificial Intelligence (GenAI) stands out for its rapid evolution and growing influence in educational environments. What began as a computational

J. A. C. Herrán (✉)
Francisco de Vitoria University, Madrid, Spain
e-mail: jorge.cerdio@ufv.es

L. A. Garcia-Segura (ed.), *Collaborative Insights on Technology and Legal Frameworks*, Advanced Sciences and Technologies for Security Applications,
https://doi.org/10.1007/978-3-032-27714-5_1

novelty is now redefining the contours of teaching and learning—especially for the youngest learners.

While many celebrate GenAI's potential to personalize instruction and support teachers, others have raised critical concerns. Educators, ethicists, and parents alike have begun to question the deeper impact these systems may have on child development, educational values, and the human relationships at the core of learning. Their voices, including my own, call for a deliberate and thoughtful assessment of this technological shift.

This work aims to contribute to that reflection. Focusing on the early childhood context, it explores what GenAI is, the risks it introduces, and how we might evaluate its coherence with the foundational aims of education. Central to this inquiry is the conviction that educational technologies must serve—not substitute—the formative vocation of education and the dignity of the child.

2 What is Educative AI for Young Children

GenAI in education can be defined as the application of advanced machine-learning models—most notably large language and multimodal models—that are capable of autonomously producing novel instructional content, assessments, and learner supports. At its most fundamental level, GenAI "refers to algorithms, such as ChatGPT, that have the capacity to create diverse content, including audio, code, images, text, simulations, and videos" (Akhtar 2024, p. 2). When integrated into teaching and learning environments—sometimes called "Educative AI" (EdAI)—these systems can generate personalized feedback, scaffold complex problem-solving tasks, and even simulate one-on-one tutoring interactions.

EdAI encompasses a diverse array of activities wherein generative models, such as ChatGPT, enhance both teaching and learning. From the user's perspective, students and educators interact with GenAI as dynamic collaborators; from the institutional perspective, EdAI provides scalable, data-driven support that transforms traditional educational practices.

EdAI is reshaping educational practice by embedding sophisticated content-creation, personalization, and analytics directly into the learning process. From the learner's perspective, students benefit from highly adaptive platforms that continuously analyze their performance to tailor difficulty, recommend resources, and even integrate personal interests and cultural contexts into curricula. Mittal et al. (2024, p. 142,737) demonstrate how EdAI systems can foster inquiry by tracking individual progress and dynamically adjusting lesson pathways to each student's needs. Simultaneously, educators leverage EdAI to automate the generation of pedagogical materials—ranging from open-ended question prompts to detailed rubrics—thereby freeing instructional designers to focus on higher-order tasks. For example, Bhat et al (2022, p. 23) show that automated question-generation research outlines methods by which EdAI produces and evaluates domain-specific assessment items.

Beyond content creation, EdAI serves as an around-the-clock tutor and formative assessor. From the user's perspective, interactive chatbots offer Socratic questioning, multilingual conversation practice, and immediate explanatory feedback, effectively acting as "24/7" virtual instructors. At the same time, institutions benefit from automated grading engines that evaluate submissions in real time—identifying misconceptions, quantifying mastery levels, and prescribing corrective exercises. BaiDoo-Anu and Owusu Ansah (2023, p. 52) highlight how ChatGPT-based systems embed continuous feedback loops into learning environments, reducing educator workload while ensuring timely, data-driven intervention.

Finally, Pellas (2025, p. 3) shows that EdAI amplifies engagement and supports advanced research and professional development. Learners interact with richly illustrated simulations, AI-generated videos, and interactive e-books that render abstract concepts—such as molecular processes or historical reconstructions—visually tangible. Concurrently, researchers and educators harness EDAI to code qualitative data, synthesize literature, and generate domain-specific exercises—such as programming problems automatically crafted to match learner proficiency, Chen et al. (2020, p. 75,270). In doing so, EdAI establishes an integrated ecosystem that personalizes instruction, automates routine tasks, and fosters both student skill development and continuous teacher learning.

Together, these activities constitute an integrated educative ecosystem that personalizes instruction, automates routine tasks, enriches content, and supports both learner development and educator expertise—pointing to a transformative future for AI in education.

Focusing solely on the student's perspective EdAI brings new challenges when applied to specific age ranges, that of young children, from children birth to eight years.

3 What are the Risks in EdAI for Young Children

The incorporation of EdAI into early educational settings presents not only opportunities but a web of multifaceted risks—particularly for young children, whose developmental, cognitive, and emotional capacities are still in formation. Unlike static digital tools, EdAI produces dynamic and interactive outputs in real-time, which, according to Yu et al. (2025, p. 2) introduces a novel category of risk that is personalized, unpredictable, and often opaque.

A primary concern is the generation of inaccurate or misleading information. Young children, particularly in early developmental stages, often lack the critical faculties necessary to discern falsehoods or inaccuracies embedded in EdAI outputs. EdAI systems, such as Bing's GenAI producing a blue sun or ChatGPT fabricating citations, may unintentionally instill incorrect mental models in children as been argued by Kosoy et al. (2024, p. 3). Furthermore, these systems are capable of generating content Not Safe For Work or otherwise inappropriate content, especially when inputs are guided by curiosity—common among young children.

Yu et al. (2025, p. 4) proposes that the real-time generation of toxic content based on innocuous queries exacerbates the risks.

Overreliance on GenAI for answers can undermine critical thinking and creative problem-solving—skills essential for lifelong learning. Picton and Clark (2024, p. 2) report that "1 in 5 admitted to just copying AI outputs, and a similar percentage said they didn't check responses from AI in case they might be wrong," highlighting a need for stronger support in developing young people's critical and creative engagement with GenAI. More alarmingly, prolonged interaction with emotionally responsive GenAI companions may foster emotional dependency, social withdrawal, or reinforce harmful behaviors like self-harm or bullying without detection or intervention. Children often treat GenAI not as tools but as peers or mentors, raising the specter of affective displacement in formative years. Hoffman et al. (2021, p. 3) found that children often form emotional parasocial relationships with conversational agents, treating them as "human-like entities with feelings and for whom they have feelings," suggesting that these agents are perceived not merely as tools, but as peers or confidants.

GenAI relies heavily on personal data to optimize outputs, raising acute concerns about privacy and data protection. As reported by Hoffman et al. (2021, p. 5) exploring parent–child perceptions of GenAI, children may disclose deeply personal information—such as trauma histories and private family details—while interacting with GenAI perceived as companions, often without understanding the privacy implications. Many educational institutions lack the infrastructural safeguards to prevent breaches or misuse of this data according to Wang et al., (2025, p. 2). Further, Yu et al. 2025, p. 7) argues that GenAI's persuasive capacities often guide children into revealing more than they intend, creating inadvertent surveillance environments.

Bias in GenAI outputs stems not only from training data but also from algorithmic design. Children exposed to these biases may internalize stereotypical representations of race, gender, and culture, which Vartiainen et al. (2024, p. 5) argues can shape their worldview in insidious ways. Most models are trained predominantly on English-language datasets, marginalizing non-native speakers and reinforcing linguistic hierarchies in global education systems according to Mezak Matijevic et al. (2024, p. 3).

The opacity of GenAI decision-making significantly complicates risk management. Parents and educators frequently struggle to comprehend how and why these systems operate as they do—a challenge intensified by inadequate transparency standards in educational deployments, as highlighted by Yan et al. (2023, p. 91). This lack of clarity hampers efforts to effectively mediate risks or hold platforms accountable when harms arise. In response, calls for Explainable GenAI in educational contexts have intensified, as noted by Giannakos et al. (2024, p. 4), aiming to make GenAI processes more interpretable and trustworthy.

Yu et al. (2025, p. 6) highlights that beyond accidental harms, GenAI also lowers the threshold for deliberate misuse. Children have used GenAI to generate violent stories, fake identities, or sexually explicit content—sometimes targeting peers. The risk of children becoming both victims and perpetrators of digital harm is heightened in these unregulated GenAI interactions. Compounding this are concerns about

GenAI-facilitated academic misconduct and deepfakes, which threaten educational integrity and student safety as stated by Yan et al. (2023, p. 92).

Overall, while EdAI holds promise, its use with young children requires careful consideration of these multifaceted risks, spanning content, development, privacy, bias, ethics, and potential misuse. It is crucial to approach its integration responsibly and ethically, with appropriate frameworks, policies, and support for educators, parents, and students.

Given the significant risks, particularly for young children with developing cognitive and emotional capacities, a potential next step could be to prioritize the development and implementation of clear guidelines and policies specifically for the use of EdAI in early childhood education settings, informed by developmental psychology and GenAI safety experts.

4 How to Approach Risks in EdAI for Young Children

While identifying the diverse risks associated with EdAI for young children is essential, an exclusively risk-based perspective may be insufficient for providing a constructive path forward. In response, we propose a new approach—*making sense of EdAI from the educational model underlying the question.* Rather than treating EdAI as a neutral tool to be judged solely by its outputs, this perspective urges stakeholders to evaluate coherence with the underlying educational model, pedagogical goals, alignment with child development principles, and the dynamic function EdAI serves within specific learning contexts. This shift reframes safety and efficacy not as add-ons, but as integral to how EdAI is designed, deployed, and understood in early childhood education.

An educational model serves as the highest-level conceptual framework within an educational institution, encapsulating its foundational beliefs regarding the purposes of education, human development, and the processes of teaching and learning. It is not merely a pedagogical instrument, but rather a philosophical and strategic declaration of institutional identity, history, values, and vision (Universidad Francisco de Vitoria 2021, pp. 9–12). According to Tünnerman Bernheim (2008, p. 15), the educational model articulates the educational paradigms that an institution professes and serves as a reference for all institutional functions—teaching, research, and community engagement. It influences curriculum design, the selection of pedagogical strategies, and the definition of student competencies, and is thus the foundational axis upon which pedagogical and didactic models rest. The model is inherently abstract and subject to evolution, shaped by sociocultural transformations and philosophical reconsiderations of what it means to educate and to be educated (Bushman 2022, pp. 5–6).

Before engaging in any risk–benefit analysis regarding the use of EdAI with young children, the foundational question must concern its coherence with the institution's educational model. This preliminary assessment is not a procedural formality but a substantive evaluation of whether the use of such technology aligns with the institution's philosophical anthropology, vision of the human person, and core pedagogical commitments.

Consider a paradigmatic case: if an educational model is grounded in a philosophical understanding of the human being as a relational, developing, and meaning-seeking subject (Universidad Francisco de Vitoria 2021, pp. 9–12). If such a model centers on personal encounter, ethical formation, and dialogic learning, then the deployment of EdAI must be assessed against these values. As Tünnerman Bernheim (2008, p. 15) emphasizes, educational models function as interpretive frameworks through which all pedagogical decisions are evaluated—including technological adoption.

According to Bushman (2022, p. 5), the anthropocentric foundations of 21st-century education necessitate a focus on human development, creativity, responsibility, and ethical maturity. If EdAI compromises the cultivation of these capacities—by displacing dialogic learning with one-sided algorithmic outputs, encouraging parasocial dependencies (Hoffman et al. 2021, p. 3), or eroding critical thinking (Picton and Clark 2024, p. 2)—then it is prima facie incoherent with an educational model predicated on the development of the whole person. Moreover, in models that espouse a personalist or community-oriented anthropology, the displacement of human educators by algorithmic interlocutors may undermine core relational dynamics foundational to early childhood education.

Thus, before assessing whether EDAI poses developmental, cognitive, or ethical risks—which it clearly does (Yu et al. 2025, pp. 2–7)—educators must first ask: does EdAI reinforce or contradict our institutional understanding of education as the formation of persons in community? Does it facilitate or fracture the pedagogical relationship understood not merely as content transmission but as personal accompaniment? If the answers are negative, the deployment of EdAI in early childhood settings may not only be imprudent but structurally misaligned with the institution's deepest commitments. In short, the coherence of EdAI with the educational model is the precondition for any legitimate consideration of its risks or benefits.

5 Assessing the Coherence of EdAI Use in Early Childhood: Guiding Principles for Educational Models

Philosophy and philosophical anthropology serve as the foundational axis from which educational models derive their purpose, coherence, and legitimacy. Educational philosophy provides the normative grounding for educational aims, methods, and values—it determines not only what education ought to accomplish but also the criteria by which it can be evaluated (Bushman 2022, pp. 5–6). Philosophical anthropology clarifies what it means to be human and, in doing so, delimits what is educationally desirable and possible. It provides the ontological and moral parameters for the formation of the learner, emphasizing the importance of "human-dimensional constants" that anchor educational practice in stable anthropological insights rather than ephemeral technical trends (Bushman 2022, p. 6). Without these philosophical underpinnings, educational initiatives risk becoming "shaky," directionless, and susceptible to technocratic instrumentalization.

The incorporation of new technologies, particularly Artificial Intelligence, must be assessed against these foundational principles. As argued by Santoianni (2014, pp. 1044–1046), technological interventions in education are not neutral tools but deeply transformative forces that reshape educational relationships, knowledge production, and even learner subjectivity. Bushman (2022, p. 6) emphasizes that in the twenty-first century, the human-centric character of education is increasingly under pressure from efficiency-driven, technocratic paradigms. In such a context, failing to evaluate the use of EdAI through the lens of philosophical anthropology risks adopting tools that undermine the very conception of the learner as a relational, moral, and meaning-making subject. Rather than serving human development, AI may inadvertently supplant or distort it. Thus, a philosophically anchored evaluation—sensitive to questions of human dignity, freedom, and formation—is essential not only for assessing technological compatibility, but also for preserving education's role as a cultural and civilizational cornerstone (Bushman 2022, p. 5).

To preserve this coherence in the age of EdAI, especially within early childhood education, it is imperative to engage in structured, philosophically rigorous inquiries that interrogate the anthropological adequacy, ethical orientation, and relational effects of such technologies. These inquiries must transcend utilitarian calculations of efficiency and instead ask whether and how EdAI contributes to the integral formation of the human person.

First, a foundational question arises: What conception of the child is embedded in EdAI systems used in early education? If the underlying model treats the child primarily as a recipient of data or as a behaviorally modifiable subject, this betrays a reductive anthropology that is incompatible with both classical educational thought and contemporary philosophical insight (Papastephanou 2014, p. 160; Fleury and Garrison 2014, pp. 24–26). The child, as philosophical anthropology reminds us, is not merely a learner but a becoming being—a nexus of potentiality, relation, and narrative. An EdAI application that fails to recognize this ontological depth risks instrumentalizing childhood and flattening the human developmental arc into computational metrics.

Second, one must examine the teleology implicit in the design and application of EdAI. That is, toward what ends is this technology directed? Tools that are solely optimized for measurable outcomes—test performance, behavioral compliance, or data extraction—may be in profound tension with the deeper aims of education: the cultivation of wisdom, ethical judgment, and the capacity for contemplative thought (Pius XI 1929, §6; Bushman 2022, pp. 6–7). If EdAI cannot support, or worse, obstructs these ends, its educational legitimacy must be called into question. Technology should not dictate the ends of education but must remain subordinate to them.

Third, the methodological coherence between EdAI and developmental appropriateness must be critically scrutinized. In early childhood, where cognitive, emotional, and social domains are inextricably intertwined, pedagogical approaches must respect the child's holistic rhythm. Does the EdAI tool align with the dialogical, exploratory, and play-based learning essential to early development, or does it impose rigid sequences and premature abstraction? As Cheng (2023, p. 2285) and Santoianni (2014, p. 1045) argue, the value of a model lies not in its technological sophistication, but in its consonance with the anthropological and epistemological needs of the learner.

Fourth, any philosophical evaluation of EdAI must address its impact on the relational fabric of education. Learning, particularly in early childhood, is an embodied, affective, and fundamentally interpersonal process. Catholic pedagogical traditions emphasize the communal character of education, viewing it as a shared journey animated by charity, dialogue, and witness (*Gravissimum Educationis* 1965, §1; Francis 2015, §9). If AI systems mediate or displace human relationships, particularly those between educator and child, the integrity of the learning encounter is compromised. The risk is not merely pedagogical inefficacy but anthropological distortion—eroding the child's capacity for empathy, trust, and dialogical self-understanding.

Fifth, ethical oversight must be integral to any deployment of EdAI. The Vatican's 2024 "Linee Guida in materia di intelligenza artificiale" insists that AI must remain at the service of the person and be governed by principles of transparency, justice, and respect for human dignity (Pontifical Commission 2024, §§2–3). These guidelines are especially crucial in early childhood, where the learner's autonomy, privacy, and moral formation are in their most vulnerable and formative stages. The ethical evaluation of EdAI thus requires attention not only to outcomes but to the underlying value systems encoded in its architecture and operations.

Finally, the criteria by which EdAI's educational impact is assessed must reflect a holistic vision of learning. If success is measured solely in terms of quantifiable performance data, we risk succumbing to an epistemological reductionism that marginalizes ethical, aesthetic, and spiritual dimensions of growth (Santoianni 2014, p. 1046; Bushman 2022, p. 7). Evaluation must itself be an expression of philosophical anthropology: it must ask not only "What did the child do?" but "Who is the child becoming?" and "Is this becoming consonant with her deepest vocation as a human person?"

Thus, a coherent EdAI implementation in early childhood education is not merely a technical matter—it is an anthropological and moral imperative. Only by anchoring its use in a reflective, principled, and relational vision of the human person can EdAI serve, rather than subvert, the transformative vocation of education.

Moreover, the early childhood context intensifies these philosophical stakes. Young children's neurocognitive, emotional, and moral development occurs within profoundly relational frameworks, often mediated through play, affective attunement, and embodied pedagogy. Any educational tool introduced at this formative stage must therefore not only be age-appropriate but existentially coherent with how children experience meaning, selfhood, and otherness (Yiu et al. 2023). EdAI that mimics human interaction without reciprocal intentionality risks confusing children's ontological boundaries, potentially fostering attachments to anthropomorphic agents that simulate but cannot participate in authentic relational dynamics (Hoffman et al. 2021, p. 4). As recent empirical studies show, children often attribute emotional and moral agency to AI entities, which could reconfigure their expectations of human interaction and ethical responsibility (Kosoy et al. 2024).

Furthermore, introducing EdAI without adequate scaffolding may subtly privilege certain epistemologies—namely those based on information retrieval, predictive analytics, and procedural correctness—at the expense of exploratory, dialogic, and imaginative modes of learning foundational to early childhood education (Vartiainen et al. 2024). This epistemological narrowing risks marginalizing the very cognitive and emotional capacities—curiosity, empathy, improvisation—that the educational model may be designed to cultivate. Without alignment to the educational telos, even sophisticated AI tools can inadvertently domesticate rather than liberate the learner's potential (Kanders et al. 2024).

In light of this, coherence assessments should operate at both macro and micro levels. At the macro level, institutions must assess whether the ontological assumptions of EdAI tools align with their own anthropological commitments. At the micro level, specific implementations should be evaluated for how they impact the lived experience of the child: Do they foster wonder or dependence? Dialogue or monologue? Critical agency or passive consumption? These questions, while pedagogical in form, are fundamentally ethical in substance (Yan et al. 2023; Yu et al. 2025).

6 Conclusions

As this exploration has shown, GenAI in early childhood education cannot be evaluated merely by its functional outputs or efficiency gains. Rather, it must be assessed through the lens of philosophical anthropology and the foundational educational model it engages or disrupts. In particular, any use of EdAI must begin by asking not only "What can it do?" but "What vision of the child does it presuppose, and what kind of human being does it help to form?"

The pedagogical and moral risks—from parasocial attachments to epistemic reductionism and anthropological flattening—are profound, particularly at a developmental stage where children are forming their sense of reality, relationality, and agency. As such, the deployment of GenAI is not neutral; it becomes a statement of institutional values and anthropological commitments.

The conceptual shift proposed here is a movement from instrumentalist framings of AI "risk" and "benefit" toward an axiological evaluation: does the deployment of EdAI reinforce the institution's anthropological commitments, its pedagogical telos, and its ethical orientations? This reflection opens a discursive space in which philosophical anthropology, developmental psychology, and educational praxis co-constitute our understanding of technological appropriateness.

Moreover, the ontological implications of early EdAI integration—particularly its simulation of reciprocity, epistemic authority, and affective engagement—demand rigorous scrutiny. As children form parasocial attachments and epistemic dependencies on algorithmic interlocutors, we confront not merely questions of safety but of anthropological distortion and pedagogical incoherence. These are not technical bugs to be patched but structural dissonances that must be addressed at the level of model design and institutional discernment.

Thus, educational institutions must engage in a radical re-centering: placing the child not at the mercy of algorithmic design, but at the heart of educational discernment rooted in a coherent philosophical framework. Only through this re-centering can we ensure that technology serves the integral formation of persons rather than distorting it. Coherence with the institutional educational model—especially those rooted in relational, communal, and ethical personalism—must be the precondition for any technological adoption.

This analysis thus calls for the development of pre-deployment coherence assessments and post-deployment relational audits that situate EdAI within a broader framework of formation, human dignity, and the integrity of the educational encounter. Such frameworks must be transdisciplinary, integrating normative educational theory, human-centered AI design, child rights law, and empirical developmental science.

This philosophical path invites us to integrate technology not as an educational savior but as a potential co-builder in a broader vision of human flourishing—carefully, cautiously, and always in fidelity to the child's dignity and vocation.

References

Akhtar ZB (2024) Unveiling the evolution of generative AI (GAI): a comprehensive and investigative analysis toward LLM models (2021–2024) and beyond. Journal of Electrical Systems and Inf Technol 11:22. https://doi.org/10.1186/s43067-024-00145-1

BaiDoo-Anu D, Owusu Ansah L (2023) Education in the Era of Generative Artificial Intelligence (AI): Understanding the Potential Benefits of ChatGPT in Promoting Teaching and Learning. Journal of AI 7:52–62. https://doi.org/10.61969/jai.1337500

Bhat S, Nguyen H, Moore S, et al (2022) Towards Automated Generation and Evaluation of Questions in Educational Domains. https://doi.org/10.5281/ZENODO.6853085

Bushman, I. (2022). Education in the 21st century: Philosophical foundations and principles. Futurity Philosophy, 1(2), 4–15. https://doi.org/10.57125/FP.2022.06.30.01

Chen L, Chen P, Lin Z (2020) Artificial Intelligence in Education: A Review. IEEE Access 8:75264–75278. https://doi.org/10.1109/ACCESS.2020.2988510

Cheng, X. (2023). A learner-centered teaching model. Sino-US International Education Forum, 15(2), 2285–2286.

Fleury, S., & Garrison, J. (2014). Toward a new philosophical anthropology of education: Fuller considerations of social constructivism. Interchange, 45(1), 19–41.https://doi.org/10.1007/s10780-014-9216-4FRn

Francis (2015) Laudato si'. Encyclical Letter, 24 May 2015. Librería Editrice Vaticana.

Giannakos M, Azevedo R, Brusilovsky P, et al (2024) The promise and challenges of generative AI in education. Behaviour & Information Technology 1–27. https://doi.org/10.1080/0144929X.2024.2394886

Hoffman, A., Owen, D., & Calvert, S.L. (2021). Parent reports of children's parasocial relationships with conversational agents: Trusted voices in children's lives. Human Behavior and Emerging Technologies, 3(1), 1–12. https://doi.org/10.1002/hbe2.271

Kanders K, Stupple-Harris L, Smith L, Gibson JL (2024) Perspectives on the impact of generative AI on early-childhood development and education. Infant and Child Development 33:e2514. https://doi.org/10.1002/icd.2514

Kosoy E, Jeong S, Sinha A, et al (2024) Children's Mental Models of Generative Visual and Text Based AI Models. arXiv:2405.13081v1.

Mezak Matijevic, M., Pisker, B., & Dokic, K. (2024). Constructing a socio-legal framework proposal for governing large language model usage and application in education. Social Sciences, 13(9), 479. https://doi.org/10.3390/socsci13090479

Mittal U, Sai S, Chamola V, Sangwan D (2024) A Comprehensive Review on Generative AI for Education. IEEE Access 12:142733–142759. https://doi.org/10.1109/ACCESS.2024.3468368

Papastephanou, M. (2014). To mould or to bring out? Human nature, anthropology and educational utopianism. Ethics and Education, 9(2), 157–175.

Paul VI. Gravissimum Educationis (Declaration on Christian Education), Libreria Editrice Vaticana, 1965.

Pellas N (2025) The Impact of AI-Generated Instructional Videos on Problem-Based Learning in Science Teacher Education. Education Sciences 15:102. https://doi.org/10.3390/educsci15010102

Picton, I., & Clark, C. (2024). Children and young people's use of generative AI to support literacy in 2024. National Literacy Trust.

Pius XI. Divini illius magistri (Encyclical on the Christian Education of Youth), Libreria Editrice Vaticana, 1929.

Pontificia Commissione per lo Stato della Città del Vaticano (2024) Decreto n. DCCII – Linee guida in materia di intelligenza artificiale. Governatorato dello Stato della Città del Vaticano. https://www.vaticanstate.va/images/N.%20DCCII.pdf Accessed 15 Jan 2025.

Santoianni, F. (2014). Models in pedagogy and education. In: Models in Engineering, Architecture, and Economical and Human Sciences. SpringerReference, pp. 1033–1048. https://doi.org/10.1007/978-3-642-27737-5_49-1

Tünnerman Bernheim, C. (2008). Modelos educativos y académicos. Managua: Hispamer.

Universidad Francisco de Vitoria. (2021). Formar para transformar en comunidad: El Proyecto Formativo de la Universidad Francisco de Vitoria. Madrid: Editorial UFV.

Vartiainen, H., Kahila, J., Tedre, M., López-Pernas, S., & Pope, N. (2024). Enhancing children's understanding of algorithmic biases in and with text-to-image generative AI. New Media & Society. https://doi.org/10.1177/14614448241252820

Wang, Y., Zhou, R., & Zhou, S. (2025). Current situation and prospect of AI education application. International Journal of Education and Social Development, 2(3), 162–175.

Yan, L., Sha, L., Zhao, L., Li, Y., Martinez-Maldonado, R., Chen, G., Li, X., Jin, Y., & Gašević, D. (2023). Practical and ethical challenges of large language models in education: A systematic scoping review. British Journal of Educational Technology, 55(1), 90–112. https://doi.org/10.1111/bjet.13370

Yu Y, Liu Y, Zhang J, et al (2025) Understanding Generative AI Risks for Youth: A Taxonomy Based on Empirical Data. https://doi.org/10.48550/ARXIV.2502.16383

European Union and Digitization: Analysis of the Application of the New Artificial Intelligence Regulation

Virginia Saldaña Ortega

Abstract This paper examines the European Union's landmark Artificial Intelligence Regulation, situating it within the broader context of the EU's Digital Strategy and pursuit of digital sovereignty. It outlines the normative Europeanization process that led to the adoption of a risk-based regulatory framework, emphasizing the transition from the GDPR's data protection paradigm to a comprehensive AI governance model. Key institutional innovations—most notably the creation of the European Artificial Intelligence Office—are analysed for their role in ensuring uniform application, supervising high-risk systems, and fostering cooperation between national authorities. The Regulation's political dimension is explored through the lens of the "Brussels Effect," illustrating the EU's capacity to export human-centered governance standards globally. The paper also addresses internal challenges, such as balancing competences between the EU and Member States, adapting national supervisory structures, and mitigating institutional tensions related to subsidiarity. Further, it examines the coexistence of AI and personal data protection, highlighting issues of automated decision-making, algorithmic bias, and the necessity of privacy-preserving technologies. Concluding remarks underscore the Regulation as both a legal instrument and a political declaration, with profound implications for the legitimacy, credibility, and strategic autonomy of the EU in the digital age.

1 Introduction

Since its inception in the 1950s, the European project has evolved into a supranational legal framework of considerable complexity and ambition, capable of harmonizing state wills in pursuit of common goals. This institutional construction, born of the vision of figures such as Jean Monnet and Pierre Du Bois, was built on the pillars of progressive integration, structured cooperation, and the overcoming of the continent's historical divisions (Saldaña 2021).

V. S. Ortega (✉)
Antonio de Nebrija University, Madrid, Spain
e-mail: vsaldana@nebrija.es

L. A. Garcia-Segura (ed.), *Collaborative Insights on Technology and Legal Frameworks*, Advanced Sciences and Technologies for Security Applications,
https://doi.org/10.1007/978-3-032-27714-5_2

Today, more than seven decades later, that legacy takes on new dimensions in light of the contemporary challenges posed by artificial intelligence and digital transformation. The digitization of our societies not only redefines economic and social relationships but also demands a shared ethical, legal, and political framework that safeguards fundamental rights and ensures the responsible use of emerging technologies.

In this context, integration and legislative harmonization are no longer mere political ideals, but structural necessities. The European Artificial Intelligence Regulation is a prime example of this integrative vocation, proposing a coordinated and coherent response to challenges that transcend national borders. Regulating AI requires unified criteria, shared capacities, and a common vision that only a cohesive European Union can provide.

Thus, in times of political uncertainty, geostrategic tensions, and accelerated technological transformations, the commitment to a digital, secure, and humanist Europe becomes the most solid path to strengthening the sense of belonging, institutional solidarity, and the international relevance of the European model. It is not merely about adapting to change, but about leading it, guided by the values that gave birth to this community of nations.

Artificial intelligence (AI) has become one of the most relevant topics of our time. In recent decades, we have witnessed remarkable advances in process automation at all levels—an automation that has created a true web of activity in which digitization and AI are deeply embedded, significantly facilitating and simplifying procedures across many areas of our reality (Veale 2021).

Addressing a subject of such complexity, which involves multiple disciplines and is advancing at a dizzying pace, represents a challenge for any researcher. It is not unlikely that some of the positions presented here may soon become outdated due to the sector's rapid evolution. The very dynamic nature of AI generates constantly shifting scenarios that require continuous review of its legal, social, and economic implications.

Artificial intelligence can be considered one of the most significant technological milestones of the twenty-first century, comparable in impact to the invention of the printing press or the emergence of the Internet. This tool has the potential to influence how we think, work, and organize ourselves as a society, generating fundamental debates in both scientific and legal domains. Around it coexists optimistic perspectives—highlighting its utility and transformative capacity—and more critical views that warn of its risks.

Certain alarmist narratives, especially those promoted by sensationalist media, have contributed to a distorted perception of AI, portraying it as an existential threat. In contrast to these discourses, this work adopts a constructive perspective, recognizing the usefulness of AI as a tool at the service of the common good (Saldaña 2025).

Given the cross-cutting interest that this technology arouses, it is appropriate to pose a basic yet fundamental question: what do we mean by artificial intelligence? Generally speaking, it can be defined as a set of tools or systems capable of processing large volumes of data, mimicking human cognitive processes in tasks such as content generation, decision-making, and outcome prediction. However, this definition varies depending on the disciplinary lens through which it is analysed.

In the institutional field, for example, the European Parliament has adopted a simple and synthetic definition, stating that:

Artificial intelligence is the ability of a machine to exhibit the same capabilities as human beings, such as reasoning, learning, creativity, and planning.

Regardless of the precise definition adopted, the close link between artificial intelligence and law is unquestionable, given its growing influence on legal interpretation, outcome prediction, and the management of complex processes.

2 The Normative Europeanization Process of Artificial Intelligence

Just as the steam engine or electricity once did in the past, artificial intelligence is profoundly transforming our society and industry. Thanks to the exponential increase in computing power, the massive availability of data, and the evolution of algorithms, AI has become a first-order strategic technology. Its regulation, therefore, is an unavoidable necessity (Saldaña 2025).

The European Union has played a prominent role in building a common regulatory framework that enables the safe and efficient deployment of these technologies. Far from merely reacting, the EU has taken a proactive stance in designing legal standards for the integration of artificial intelligence into everyday life, positioning itself as an international benchmark in the field.

The General Data Protection Regulation (GDPR) already demonstrated the EU's ability to harmonize fundamental rights and principles in complex technological contexts. In the same vein, the recent adoption of the Artificial Intelligence Regulation reaffirms this commitment, placing the Union at the forefront of global digital governance.

The new regulation is based on a risk-based approach, recognizing that the very characteristics that make AI valuable may also pose threats to security, fundamental rights, and democratic values. As a result, the regulation establishes a normative differentiation according to the level of risk posed by each AI system, ranging from minimal risk to unacceptable risk, the latter being strictly prohibited (Floridi 2023).

The European legislator opted for a regulatory instrument to ensure uniform application across all Member States. As stated in the preamble of the text, the aim is "to apply the new rules in a homogeneous manner" and to ensure that the economic and social benefits of AI are not overshadowed by its potential adverse effects.

This legal approach is not isolated, but rather part of a broader trend that some experts have referred to as the "Brussels Effect": the EU's ability to export its regulatory standards beyond its borders. Just as occurred with the GDPR, the Artificial Intelligence Regulation is expected to influence the legislation of third countries, especially those with close commercial ties to Europe (Saldaña 2024).

The globalization of European values through the inclusion of regulatory clauses in free trade agreements or bilateral treaties could lead to the progressive adoption of the regulation's principles by non-European companies. This will be particularly evident in sectors such as public procurement, where economic operators will seek to comply with a demanding regulatory framework to maintain competitiveness in the market.

Several months after its publication, a fundamental question arises: has the Artificial Intelligence Regulation met expectations? Asking such a broad question inevitably gives rise to varied and diverse responses, and indeed, the answer is far from unequivocal. From a normative standpoint, the AI Regulation must be considered a truly pioneering step, as it represents the first serious global attempt to comprehensively regulate this technology through a risk-based approach that prioritizes fundamental rights, transparency, and human oversight. In this sense, its symbolic and structural value is clear and already resonates beyond the European context.

If we look at the question from a practical perspective, it is not easy to provide a definitive answer, as its real value is still in a testing phase. Thus, the effectiveness of the regulation will largely depend on its concrete implementation, on the ability of Member States to provide sufficient resources to the competent authorities, and on the agility to update technical criteria in a constantly evolving innovation environment.

And this is precisely the issue that is currently raising the most concerns: is its real-world application useful and effective in the whirlwind of digital change and evolution that seems to overwhelm us every day? Although the regulation offers a clear legal structure, the potential hindrance to innovation caused by excessive bureaucracy is one of the main concerns today. Therefore, while the regulation provides a valuable starting point as a declaration of principles and a basic standard for the protection of individual rights, only time and practical implementation will determine the regulation's true effectiveness, efficiency, and its ability to adapt and evolve in line with the demands of digital transformation—year by year, month by month, or even day by day.

Moreover, there are other substantive and operational issues that have generated significant criticism in various sectors during the short time since the regulation's publication in the Official Journal of the European Union. The first issue worth highlighting is the excessive complexity and vagueness of the regulation—a recurring problem in EU legislative and regulatory development. In this case, the final text is long and technically dense, which makes it difficult for developers and users of AI systems to interpret and apply correctly. Similarly, the ambiguity of certain concepts creates areas of uncertain interpretation and opens the door to conflicting understandings of the text.

Another major issue that scholars have highlighted over time is the mismatch between regulation and the speed at which technology advances. As mentioned earlier, the rigid nature of positive law—particularly in the case of regulations of general application—can hinder the legal framework's ability to adapt to rapid technological changes. A clear example of this lack of agility and adaptability can be seen in the AI regulatory process itself, whose initial proposal dates back to 2021, with the final regulation only being adopted in 2024.

This issue leads us to examine one of the most debated aspects in recent legal literature: the choice of a regulation as the legal instrument for governing artificial intelligence in the European Union. Evidently, this choice was neither trivial nor accidental on the part of the legislator. It responds to the need to ensure direct, homogeneous, and binding application across all Member States, without the need for national transposition. The aim was to avoid fragmentation of the European digital market and to guarantee a uniform level of protection of fundamental rights in the face of AI-related risks.

However, the core of the criticism against choosing this legal form lies in the fact that, although application is direct, effective implementation is neither automatic nor free from difficulties. Member States must create or adapt competent national authorities, establish supervision mechanisms, and coordinate with the future European Artificial Intelligence Office to ensure proper safeguards and to achieve the general objectives of correct enforcement. These tasks have already revealed the disparities and deficiencies that exist between different EU Member States.

Would it have been more appropriate to adopt a directive on AI? While this option would have allowed Member States to tailor the regulation's principles to their specific contexts, it would also have run the risk of facing the usual problems associated with directive transposition and subsequent non-compliance with EU law. Moreover, it would have sacrificed regulatory uniformity in favor of greater national flexibility—leading potentially to fragmented legislation, legal uncertainty, and unequal protection of citizens' rights.

Therefore, while the regulation as a European legal instrument represents an ambitious commitment to legal cohesion, its success will depend not only on the nature of the act itself, but on the actions undertaken to ensure its compliance. This requires coordinated implementation, appropriate technical support from European institutions, and the development of harmonized interpretation mechanisms. In short, while we are faced with an instrument that is coherent with its intended goals, its effectiveness depends on a sustained political effort to translate formal equality into real and equitable outcomes across all EU Member States (Saldaña 2024).

3 Challenges for the Coexistence of Artificial Intelligence and Personal Data Protection

Artificial intelligence is a rapidly growing tool. Its use increases day by day, as humans have come to understand that its implementation can yield significant benefits—whether in professional activities or in personal and family contexts. Therefore, it is essential to recognize that, while artificial intelligence offers new opportunities, it also generates new challenges that must be addressed and resolved appropriately and coherently.

The first of these challenges relates to decision automation. Today's AI systems are capable of making automatic decisions that directly affect individuals and their inherent rights. Clear examples include the granting of loans, candidate selection in recruitment processes, or the awarding of contracts in public procurement procedures.

The second major issue that the development of artificial intelligence poses for personal data protection concerns algorithmic bias. AI bias is rooted in the data with which these systems have been trained. Such biases can lead to discrimination based on factors such as gender, race, or sexual orientation—posing significant challenges by undermining principles of justice and fairness, as well as other core values that deeply impact an individual's life.

The European Union's General Data Protection Regulation (GDPR), enacted in 2016, has served as an essential foundation and a necessary protection standard for the development of regulations relating to artificial intelligence. It ensures that the advancement and implementation of these digital tools is carried out ethically, responsibly, and in full respect of fundamental rights.

As previously mentioned, artificial intelligence is now considered one of the most innovative drivers of data-based digital transformation. And it is precisely this data—powering the predictive and adaptive capabilities of AI algorithms—that can create serious issues in the normal course of activity. From medical data to consumer habits, online preferences, and biometric information, these represent some of the most personal and sensitive forms of data at stake. Consequently, personal data protection sits at the core of ethical and legal reflection for jurists.

These circumstances underscore the absolute necessity of designing mechanisms aimed at safeguarding and guaranteeing these privacy rights—without hindering technological progress. That is the challenge. The application of differential privacy techniques, encryption, or the creation of secure execution environments are just some of the tools developed to maintain high standards of security and protection for individual rights, especially regarding personal data.

Indeed, the symbiosis between artificial intelligence and personal data protection is both necessary and fundamental to the proper functioning of mechanisms that safeguard human rights. This symbiosis must not be relegated to a mere bureaucratic obligation, but must be understood as an essential duty, embedded in a dynamic process of change in which the development of intelligent systems evolves in parallel with the consolidation of robust organizational and legal frameworks. Only in this way will it be possible to harness the enormous potential of artificial intelligence without compromising individuals' ownership and effective control over their data—an indispensable foundation for building and strengthening fairer, safer, and more privacy-respecting digital societies.

Personal data protection in the context of artificial intelligence is a fundamental issue, and it requires a balanced approach. Both regulatory frameworks must coexist within a common reality and provide a robust normative and protective system to ensure that innovation does not come at the expense of individuals' fundamental rights. In this regard, companies, organizations, and public administrations that develop and use artificial intelligence in their daily operations must be transparent in their use of personal data, must respect user privacy, and must adhere to the established rules to prevent abuse and protect public trust in these technologies.

4 Conclusions

If the purpose of this text were to analyse the evolution of everything we know today, and we were to embark on a journey through the major milestones of the past centuries, few would dispute the historical significance of inventions such as the radio or the automobile in the twentieth century. With equal conviction, the following must be stated: digitalization and the emergence of artificial intelligence tools represent an unprecedented transformation in the recent history of our society. This transformation has its roots in the early efforts to build a technological model that gradually left behind analogy standards. This process began in the late 1980s with the creation of the World Wide Web.

There is no doubt that we are facing a tool that will play a central role in our daily lives, with tangible applications in nearly every area of social life. Artificial intelligence is poised to occupy a strategic position in the development and progress of our communities and economies.

As artificial intelligence has increasingly become part of daily life—through process automation and data-driven decision-making—the urgent need has arisen to address fundamental questions about its impact on higher dimensions, such as human rights, privacy, and essential legal principles like accountability and equal treatment.

For a long time, the subject at hand—artificial intelligence—has posed a constant challenge for our society, in a global context marked by complexity and uncertainty. This is a context in which significant challenges remain for the effective integration of these tools into existing legal and economic frameworks, largely due to one of their most notable features: variability.

The future of the Artificial Intelligence Regulation is uncertain—especially when we understand that its nature is dynamic and that regulation in this field must not be rigid or inflexible. The legal framework that governs digitalization must be characterized by ongoing evolution, facing multiple challenges in terms of implementation, harmonization, and revision, as the practice of digitalization unfolds, and artificial intelligence continues to grow and change.

One of the greatest challenges facing this Regulation is precisely the need to avoid regulatory obsolescence. The Artificial Intelligence Regulation includes mechanisms for updates, particularly concerning the annexes that list prohibited or high-risk technologies. Furthermore, closely related to this issue is the challenge of formal adoption and effective implementation. After the adoption of this Regulation at the end of 2024, its practical application in the Member States requires a transitional process involving the establishment of new instruments, the creation of national competent authorities, supervisory systems, and coordination mechanisms at the European level—elements that are still in development or trial phases. Only time will tell whether these have proven to be optimal or inadequate.

Undoubtedly, the European Union has shown its ability to lead in regulatory matters and has succeeded in positioning itself as a key factor in the push for the regulation of artificial intelligence and the protection of individuals' fundamental rights. This milestone should be seen yet another affirmation of the supranational organization's commitment to its long-standing process of integration and harmonization—a process that began more than seventy years ago and now seeks, and must continue to seek, to position the EU as a reference actor at all levels on the international stage.

References

European Parliament and Council of the European Union (2016) *Regulation (EU) 2016/679 of the European Parliament and of the Council of 27 April 2016 on the protection of natural persons with regard to the processing of personal data and on the free movement of such data (General Data Protection Regulation).* Off J Eur Union L119:1–88. Disponible en: https://eur-lex.europa.eu/legal-content/EN/TXT/?uri=CELEX%3A32016R0679

European Parliament and Council of the European Union (2024) *Regulation (EU) 2024/1689 of the European Parliament and of the Council of 13 March 2024 laying down harmonised rules on artificial intelligence (Artificial Intelligence Act) and amending certain Union legislative acts.* Off J Eur Union L168:1–165. Disponible en: https://eur-lex.europa.eu/legal-content/EN/TXT/?uri=CELEX:32024R1689

European Union (2007) *Treaty of Lisbon amending the Treaty on European Union and the Treaty establishing the European Community, signed at Lisbon, 13 December 2007.* Off J Eur Union C306:1–271. Disponible en: https://eur-lex.europa.eu/legal-content/EN/TXT/?uri=CELEX%3A12007L%2FTXT

Floridi L (2023) The European legislation on artificial intelligence: A brief analysis of the proposed AI Act. Philos Technol 36(1)

Saldaña Ortega V (2021) Análisis de los retos y desafíos futuros de la Unión Europea: El impulso federal como instrumento ejecutor del cambio. Rev Filos Juríd Soc (38):239–241

Saldaña Ortega V (2022) *El Derecho administrativo europeo.* Aranzadi, Madrid

Saldaña Ortega V (2020) Origen y evolución de la Unión Europea: Hacia la construcción de un modelo federal. In: Molina del Pozo PC (ed) *Derecho de la Unión Europea e integración regional. Libro homenaje al Prof. Dr. Carlos Molina del Pozo.* Tirant lo Blanch, Valencia, pp 115–130

Saldaña Ortega V (2024) La inteligencia artificial y su afectación en el Derecho administrativo: especial referencia a la contratación pública. Rev Unión Eur Aranzadi (4)

Saldaña Ortega V (2025) La afectación de la inteligencia artificial a los procesos de contratación pública europea. In: Asociación Española de Profesores de Derecho Administrativo (ed) *El Derecho Administrativo en la era de la inteligencia artificial: Actas del XVIII Congreso.* Aranzadi, Madrid

Veale M, Borgesius FZ (2021) Demystifying the Draft EU Artificial Intelligence Act. Comput Law Rev Int 22(4):97–112

European Union AI Act: First Regulation on Artificial Intelligence. Main Aspects of the Regulation

Jordi Regí Rodríguez

Abstract The European Union's Artificial Intelligence Act (AI Act) is a pioneering regulation aimed at overseeing AI systems within the EU. It seeks to balance AI innovation with the protection of fundamental rights, though several key issues have emerged. One primary concern is the Act's risk-based approach, which classifies AI systems according to risk level. Critics have raised concerns about the criteria for assessment and the potential risks of both over-regulating and under-regulating certain AI applications. The Act's extraterritorial application, which affects AI systems from third countries, also raises concerns about potential conflicts with international regulations. Data governance and privacy are central to the Act, which aims to protect personal data. However, there are concerns over data monopolization and the sufficiency of governance frameworks. New obligations for transparency, oversight, and risk management—particularly for small and medium-sized enterprises (SMEs)—have sparked debate about their feasibility. Additionally, questions have been raised regarding the effectiveness of enforcement mechanisms and the potential for inconsistent regulation across EU member states. Overall, the AI Act represents a crucial step toward regulating AI but highlights the challenges of addressing innovation, rights, and ethics in a rapidly advancing field. Key contentious aspects include risk classification, extraterritoriality, enforcement, and the potential for regulatory overreach.

Keywords AI Act · European Union · Artificial Intelligence

1 Introduction

Artificial intelligence (AI) has become one of the most transformative technologies of the twenty-first century, holding immense potential to revolutionize key sectors such as healthcare, education, industry, and public services (Commission 2021). As AI

J. Regí Rodríguez (✉)
EU and Public International Law, Francisco de Vitoria University, Madrid, Spain
e-mail: jordi.regi@ufv.es

L. A. Garcia-Segura (ed.), *Collaborative Insights on Technology and Legal Frameworks*, Advanced Sciences and Technologies for Security Applications,
https://doi.org/10.1007/978-3-032-27714-5_3

integrates increasingly into daily life and decision-making, establishing a regulatory framework is essential to safeguard fundamental rights and promote technological innovation ethically and safely.

In this regard, the European Union (EU) has taken a pioneering step by proposing the Artificial Intelligence Regulation, also known as the AI Act, the first comprehensive global regulation aimed at governing the use, development, and commercialization of AI systems (Parliament 2023). This regulation adopts a risk-based approach, classifying AI systems according to their potential impact on security and fundamental rights. Its primary objective is to address both the opportunities and challenges that AI presents, ensuring that its deployment is conducted responsibly.

However, the regulation has sparked some controversy. Various sectors have expressed concerns about the criteria for risk classification, arguing that they could lead to over-regulation, which might stifle innovation, or to under-regulation, which may fail to adequately protect citizens (Veale and Zuiderveen Borgesius 2021). The AI Act seeks to mitigate the ethical concerns and potential risks associated with AI while fostering innovation and ensuring the reliability of these technologies (Dignum 2023; Huang et al. 2021).

The lack of a clear regulatory framework could lead to issues such as discrimination, privacy invasions, and lack of transparency, which may harm both individuals and society as a whole (Wulf and Seizov 2020). For this reason, regulations for AI are essential to protect fundamental rights, ensure safe technological development, and foster trust in AI's applications.

The AI Act introduces a risk-based categorization system, dividing AI systems into three main categories: unacceptable risk, high risk, and minimal or no risk. Systems classified as having an unacceptable risk, such as those designed to manipulate human behavior or infringe upon fundamental rights, are prohibited under the regulation (Dignum 2023). High-risk AI systems—those used in critical areas such as healthcare, law enforcement, and financial services—are subject to stringent requirements for safety, transparency, and accountability (Dignum 2023).

Interestingly, a recent study estimates that only about 30% of AI products and projects in Germany would fall into the high-risk category under the AI Act, with the majority classified as minimal or no risk (Hauer et al. 2023). This data highlights the regulation's balanced approach, which targets only the most impactful AI applications, allowing lower-risk AI to innovate freely.

The regulation also emphasizes harmonized standards, which companies can leverage to demonstrate compliance with the AI Act's requirements. This harmonized approach seeks to ensure that a balanced regulatory framework supports innovation while also protecting EU citizens and organizations (Tartaro et al. 2023). By aligning with consistent standards, the Act enables more seamless compliance across industries, fostering an environment of both accountability and flexibility.

A significant component of the AI Act is its focus on robustness and explainability—two critical attributes for trustworthy AI systems. Technical requirements for robustness and explainability aim to establish reliable AI that can be safely used across various applications (Hamon et al. 2020). These requirements promote AI that not only performs effectively but also provides understandable, traceable outputs that increase user trust.

A thoughtfully designed regulatory framework protects users while encouraging respectable innovation. Clear rules and standards for safety and transparency create a stable environment in which businesses feel confident investing in responsible and safe AI. This regulatory clarity benefits companies by reducing risk, enhancing user trust, and boosting economic growth, as the global marketplace increasingly values ethical AI technologies (Busuioc et al. 2023).

Overall, the AI Act represents a critical step in regulating emerging technologies, balancing innovation with the need for ethical standards. Through a risk-based approach, harmonized standards, and a commitment to transparency and robustness, the regulation seeks to ensure that AI continues to be a force for positive change in a responsible and trusted manner.

2 General Framework of the Artificial Intelligence Act

The development of the EU's Artificial Intelligence Act, commonly referred to as the AI Act, emerged in response to increasing concerns about the societal and ethical implications of AI on fundamental rights. Over recent decades, the evolution of artificial intelligence has been extraordinary, with applications extending from virtual assistants to complex autonomous systems in healthcare, finance, transportation, and beyond. These advancements have sparked excitement about AI's potential but also concerns about its unintended consequences, such as bias, privacy violations, and lack of accountability (Commission 2021). Recognizing both the transformative potential and risks of AI, the EU has aimed to establish a regulatory framework that balances innovation with the safeguarding of citizens' rights.

In this context, regulation can catalyze ethical innovation, promoting the creation of AI systems that prioritize user safety, transparency, and accountability. AI developed in a regulated environment is more likely to gain public trust, fostering its acceptance and supporting its potential for positive societal impact. Moreover, a regulatory framework like the AI Act enables EU companies to lead in the development of ethical and responsible technology, potentially positioning them as competitive players on a global scale where trustworthiness and compliance are increasingly valued (Commission 2021).

The AI Act's primary objective is to mitigate the potential risks posed by AI while maximizing its benefits for society. To achieve this, the regulation introduces a balanced approach that fosters technological innovation while ensuring fundamental rights protection. Additionally, the AI Act seeks to enhance the EU's position as a leader in the development of ethical AI technologies, setting a global standard and encouraging other nations to adopt similar regulatory practices. This unified framework could facilitate a common international approach to AI, benefiting cross-border collaboration and creating a level playing field for AI development and commercialization.

The mentioned risk-based categorization system helps ensure safety and ethics in high-risk systems, the Regulation sets out specific compliance requirements. AI developers must ensure that their systems meet high standards of safety and transparency, including documentation of the system's operation, the quality of data used for training, and human oversight in case of failures or critical decisions. In addition, high-risk AI systems must be auditable and explainable, allowing users and authorities to understand the system's decision-making process.

2.1 *Historical Background and Development of the Regulation*

The creation of the AI Act reflects the European Union's commitment to regulating emerging technologies and ensuring a safe, equitable digital environment. This initiative builds on earlier efforts, notably the European Strategy for Artificial Intelligence introduced by the European Commission in 2018. The Strategy laid the groundwork for a coordinated approach to AI research, investment, and regulation across the EU (Commission 2018). Recognizing AI as a crucial driver for Europe's economic and social future, the document proposed an action plan combining technological advancement with ethical governance (Butcher and Beridze 2019).

The 2018 Strategy was further developed in 2020 with the publication of the White Paper on Artificial Intelligence, which proposed a regulatory framework centered on trust and excellence in AI (Commission 2020). This document was instrumental in shaping the AI Act, particularly through its introduction of risk-based regulation—a concept that would become central to the EU's approach to AI governance (Smuha 2021). The White Paper also highlighted the importance of international collaboration and the need for harmonized global standards to prevent regulatory fragmentation that could stifle innovation [(Veale and Zuiderveen Borgesius 2021)].

Following extensive consultations and deliberations with member states, technology experts, civil society organizations, and businesses, the European Commission formally proposed the AI Act in April 2021 (López-Tarruella 2023). This legislative proposal aimed to create a flexible legal framework capable of keeping pace with the rapid evolution of AI technologies, while grounding the regulation in the EU's fundamental values—human dignity, privacy, and non-discrimination (Floridi 2021).

On July 12, 2024, the AI Act was published in the Official Journal of the European Union and entered into force on August 1, 2024, under Regulation (EU) 2024/1689.

Official Deadlines and Phased Implementation

The AI Act is designed to roll out in stages, with specific deadlines for different chapters and requirements:

1. **Full Application**: The AI Act will be fully applicable starting August 2, 2026, with exceptions for high-risk AI uses, which are subject to accelerated timelines.

2. **Chapters I and II (General Provisions and Prohibited AI Practices)**: These provisions will apply starting on February 2, 2025, addressing general definitions, scope, and specific AI practices prohibited within the EU.
3. **Key Compliance and Oversight Chapters**: Chapters such as Chapter III, Sect. 4 (Notifying Authorities and Notified Bodies), Chapter V (General Purpose AI Models), Chapter VII (Governance), Chapter XII (Penalties), and Article 78 (Confidentiality) will come into effect on August 2, 2025, excluding Article 101, which pertains to fines for providers of general-purpose AI models.
4. **Specific Obligations**: Article 6(1) and its associated regulatory obligations will take effect on August 2, 2027, granting additional time for entities to comply with particular high-risk classification requirements.
5. **Member State Compliance Bodies**: By November 2, 2024, each EU Member State must designate public authorities responsible for overseeing and enforcing respect for fundamental rights, including non-discrimination, concerning high-risk AI systems.
6. **Guidelines from the European Commission**: By February 2, 2026, the European Commission will issue guidelines on the practical implementation of Article 6 (Classification Rules for High-Risk AI Systems) in line with Article 96. This guidance will include detailed examples of high-risk and non-high-risk AI use cases to assist with compliance.

The AI Act represents a groundbreaking regulatory framework that balances the need for innovation with safeguards for ethical AI development. Through its phased implementation, the Act aims to provide clarity and support to businesses and regulators, ensuring that AI development aligns with EU values while remaining competitive on a global scale.

Source https://www.cooley.com

2.2 Development Process and Public Consultations

The process of developing the Regulation was characterized by an inclusive and participatory approach. The European Commission conducted extensive public consultations to gather input from a wide range of stakeholders, from large technology companies to SMEs, non-governmental organizations, academics, and citizens (Parliament 2023). These consultations were essential to identify the main concerns and challenges associated with AI regulation, as well as to ensure that the resulting regulation was balanced and effective (Wong 2021).

Artificial intelligence has shown transformative potential, but it also poses ethical, privacy, and security challenges. Faced with the growing impact of AI in multiple sectors, the European Union has embarked on a process to establish a regulatory framework to ensure the safe and ethical development of this technology. This process culminated in the Artificial Intelligence Regulation, commonly known as the AI Law, which was published in the EU Official Journal in July 2024. However, the creation of this legislation was not a one-off act; it was the result of a comprehensive process of public consultations, multi-stakeholder discussions, and a series of strategic guidance documents that laid the groundwork for the development of the AI Law (Commission 2021).

The path towards AI Regulation started in 2018 with the publication of the 'European Artificial Intelligence Strategy' by the European Commission. This document recognized the importance of AI for the future of Europe and laid the foundations for a coordinated approach to AI research, development, and regulation within the EU (Commission 2018). The strategy proposed an approach that combined innovation with sound ethical principles, promoting AI as a technology of strategic importance for the economic and social development of the region (Butcher and Beridze 2019).

Building on this strategy, the EU presented the 'White Paper on Artificial Intelligence' in 2020. This document set out a framework for the development of AI in Europe based on trust and excellence and proposed the creation of a regulatory framework to address the risks and opportunities associated with this technology (Commission 2020). The White Paper was instrumental in conceptualizing a risk-based approach, categorizing AI systems according to their impact on rights and security. It also underlined the importance of international cooperation and harmonization of regulations at the global level to avoid regulatory fragmentation, which could hinder innovation (Veale and Zuiderveen Borgesius 2021).

In addition to public consultations, the development of the Regulation was also influenced by the reports and recommendations of various committees and expert groups established by the Commission. These groups provided expertise on key topics, such as AI ethics, cybersecurity, and data protection, which were integrated into the provisions of the Regulation (Goodman and Flaxman 2017).

The drafting of the AI Law included an extensive public consultation, which was essential to capture a wide range of perspectives and needs. In 2020, the European Commission launched a public consultation to gather feedback on the White Paper on

Artificial Intelligence, which received more than 1200 responses from various stakeholders, including governments, technology companies, civil society organizations, academics, and citizens (Commission 2020). The consultation sought feedback on key aspects, such as risk classification, oversight, and transparency, and significant efforts were made to ensure that the voices of all parties were heard (Smuha 2021).

The public consultations proved to be a vital space for discussion, where complex issues were raised and areas of improvement for the future AI Law were identified. Participants expressed concerns about the potential impact of strict regulation on innovation, especially for small and medium-sized enterprises (SMEs), which could be disproportionately affected by compliance costs. The need for a flexible approach that would allow AI developers and operators to adapt to regulatory requirements without slowing down technological development was also discussed (Floridi 2021).

During the process of developing the AI Law, the European Commission maintained a constant dialogue with EU Member States, which played a key role in defining the regulatory framework. The Member States participated in meetings and working groups organized by the Commission, where the objectives and technical details of the regulation were discussed. The Commission worked closely with representatives of each state to ensure that the regulation reflected the diverse national realities and needs while maintaining cohesion within the European single market (López-Tarruella 2023).

In addition, the active participation of civil society organizations and human rights groups was promoted and contributed to the discussion on the ethical risks of AI. The inclusion of these organizations helped ensure that the AI Act protected fundamental rights, such as privacy and non-discrimination, by establishing safeguards against the misuse of the technology (Butcher and Beridze 2019). Human rights organizations also advocated for transparency and accountability requirements in high-risk AI systems, ensuring that the regulation prioritizes the safety of citizens (Veale and Zuiderveen Borgesius 2021).

The process of developing the AI Law also involved AI experts, academics, and representatives of the technology sector, who brought a technical and specialized approach to the discussions. The European Commission organized forums and working groups involving AI specialists, who offered their expertise on the design and operation of AI systems, as well as on the challenges and limitations in their implementation. This dialogue allowed the Commission to incorporate a practical approach to regulation, ensuring that requirements were feasible and based on the reality of technological development (Commission 2020).

The technology sector also played a crucial role, as it was concerned about the impact that over-regulation could have on European competitiveness. Large technology companies, as well as SMEs, provided their perspective on how regulation could balance risk control with the need to foster innovation. Issues such as the classification of AI systems and the importance of a flexible and adaptive approach in regulation (Floridi 2021) were also discussed.

The legislative process also involved intense negotiations between EU member states and the European institutions, such as the European Parliament and the Council of the EU. These negotiations addressed complex issues, such as the delimitation of competencies between the EU and member states, the extraterritorial application of the Regulation, and the balance between the protection of fundamental rights and the promotion of economic competitiveness (Calo 2017).

The development and public consultation process of the EU AI Law reflects an inclusive and comprehensive approach to creating innovative and complex regulations. This process involved a wide range of stakeholders, from governments to civil society, and focused on ensuring that the regulation not only protects citizens' rights but also enables businesses to operate and develop in an ethical and safe environment. The AI Act represents a pioneering model in the regulation of emerging technologies, leading the way toward global regulation that encourages the responsible development of artificial intelligence.

3 Structure of the Regulation

The EU Artificial Intelligence Regulation is a complex and detailed piece of legislation designed to address the various challenges and opportunities presented by AI. Its structure reflects a risk-based approach that classifies AI systems according to their potential impact on fundamental rights and security (Smuha 2021). The main elements of the structure and scope of the Regulation are outlined below.

The Regulation is organized into several chapters, each of which addresses different aspects of AI regulation:

1. **General Provisions**: This chapter sets out the objectives, scope, and key definitions of the Regulation. It specifies that the Regulation applies to providers and users of AI systems within the EU, as well as to AI systems developed outside the EU but placed on the market or used within the EU. Key concepts such as "artificial intelligence," "AI system," and "provider" are also defined here (Commission 2021).
2. **Risk Rating**: This chapter introduces the risk-based approach that underpins the entire Regulation. AI systems are classified into four risk categories: low, limited, high, and unacceptable. Each category is associated with different levels of regulatory obligations and requirements. Systems deemed to present an unacceptable risk are prohibited, while high-risk systems are subject to strict controls and oversight. Limited- and low-risk systems have fewer requirements but must still comply with certain basic standards (López-Tarruella 2023).
3. **Requirements for High-Risk Systems**: This chapter details specific obligations for AI systems classified as high-risk. These obligations include risk management, transparency, human oversight, accuracy, and safety, as well as conducting compliance assessments before systems can be marketed or used. Additionally, providers of these systems must maintain detailed records and documentation and cooperate with regulatory authorities (Veale and Zuiderveen Borgesius 2021).
4. **Transparency Obligations for AI Systems**: This chapter sets out transparency requirements for AI systems, particularly those that interact directly with users.

For example, users must be informed when they are interacting with an AI system and provided with clear explanations of how certain systems work, such as automated decision-making algorithms (Floridi 2021).

5. **Supervision and Enforcement**: This chapter describes supervision and enforcement mechanisms, including the designation of competent authorities in each EU member state responsible for implementing the Regulation. It also outlines sanctions and remedial measures for non-compliance, as well as appeal procedures for those affected by AI-related decisions (Parliament 2023).
6. **Final Provisions**: This final chapter addresses issues related to the implementation of the Regulation, including transitional provisions, entry into force, and periodic reviews of the Regulation (Commission 2021).

4 Scope and Main Challenges of the Regulation

The EU Artificial Intelligence Regulation has a broad scope, covering both AI systems developed within the EU and those imported or used within its territory. This broad approach seeks to ensure that all AI systems interacting with European citizens meet the same standards of security, transparency, and respect for fundamental rights (Calo 2017).

The scope of the EU AI Act covers both providers and users of AI systems within the European Union, as well as certain systems developed outside the EU that are placed on the EU market or used within the region.

Artificial intelligence (AI) has become one of the most important technologies of our time, posing both opportunities and challenges. In response to this development, the European Union (EU) has proposed an AI Regulation, which aims to establish a regulatory framework that ensures security and fundamental rights. This regulation has significant extra-territorial implications, as it will apply not only to companies within the EU but also to those outside its borders if their activities affect the European market.

4.1 Extraterritorial Application

One of the most prominent features of the Regulation is its extraterritorial application. This means that AI companies and developers outside the EU, who wish to market or use their systems in the European market, must comply with the requirements set by the regulation. This approach reflects the EU's growing influence on global regulation, similar to what happened with the General Data Protection Regulation (GDPR), which also has an extraterritorial reach (Goodman and Flaxman 2017). The intention is to avoid loopholes that can be exploited by foreign actors and to ensure a uniform level of protection for all European citizens.

The implications of extraterritoriality have different implications according to the regulation of the Act:

- Impact on innovation: The extraterritorial application of the regulation may have a significant impact on innovation, particularly for technology companies operating outside the EU. To comply with the regulation, these companies must invest in technological and administrative adaptations to ensure that their AI systems meet the standards of transparency, accountability, and explainability set by the EU. These costs may be prohibitive for some companies, especially startups or SMEs, which could lead them to limit access to their products in the European market or to avoid investing in AI technologies that require a high degree of regulatory compliance (De Bruyne and Vanleenhove 2021).

Moreover, regulation could reduce the competitiveness of non-European firms compared to European firms, which are already used to operating under strict regulatory frameworks. In the long run, this could foster a kind of 'regulatory barrier' that makes it difficult for foreign companies to enter the EU market, affecting the diversity of AI products and services available to European consumers.

- Protection of fundamental rights and security: The EU has made it clear that one of the main objectives of the regulation is to protect citizens' fundamental rights. This includes aspects such as privacy, security, and non-discrimination. The regulation imposes restrictions and requirements on AI systems according to their level of risk. For example, systems classified as 'high risk', such as those used in labor recruitment, financial services, or healthcare, will have to comply with strict requirements for transparency, human oversight, and accountability. Companies must provide documentation explaining how the system works, how data is collected and processed, and how potential negative impacts are mitigated (Birchfield 2024).

This focus on human rights and security protection can be beneficial for EU citizens, as it increases the transparency and trustworthiness of AI systems. However, it poses challenges for companies outside the EU, which must ensure that their AI systems are not only effective but also explainable and auditable to comply with EU regulations.

- Legal liabilities and sanctions: The regulation includes significant penalty provisions for those who fail to comply with the regulation. Fines can reach up to 6% of a company's global revenues, a measure that underlines the seriousness with which the EU treats compliance with this regulation. This severity of sanctions reflects the EU's commitment to protect its citizens from the potential risks of AI by taking a firm stance against companies that might ignore the standards set for entering the EU market (Czerniawski 2024).

Such sanctions also imply several legal responsibilities for companies to ensure that their AI systems are compliant from design through to implementation and ongoing monitoring. Companies, both inside and outside the EU, will need to ensure that they have the necessary procedures and controls in place to comply with the regulations and avoid sanctions.

- Challenges in implementing offshore regulation: Implementing regulation outside the EU poses logistical and legal challenges. To impose sanctions and ensure compliance, the EU relies on cooperation with authorities and regulators in other countries. The EU's ability to impose sanctions and enforce compliance extraterritorially depends to a large extent on the existence of international collaboration and mutual recognition agreements. Such collaboration can be complex, as some countries may be unwilling to accept extraterritorial enforcement of EU rules, which may make it difficult to impose sanctions on companies operating in those jurisdictions (Benini 2022).
- Ethical challenges and regulatory dilemmas: The extraterritorial application of the regulation also raises ethical questions about the sovereignty and autonomy of other countries to set their own AI regulations. While the EU argues that its regulation protects the fundamental rights of its citizens, it also faces the challenge of balancing this approach concerning other countries' regulations and regulatory frameworks. This kind of extraterritoriality could lead to diplomatic tensions with countries that perceive EU regulation as an encroachment on their ability to regulate their technology industries.

The extraterritorial regulation of the EU AI Regulation reflects the European bloc's commitment to protecting rights and promoting transparency in the use of artificial intelligence. This regulation not only establishes a robust regulatory framework for AI within the EU but also sets an important precedent for other countries, which could adopt similar regulations in the future. However, the extraterritorial application presents significant challenges in terms of implementation, compliance costs, and potential diplomatic tensions. As the EU moves forward with the implementation of this regulation, it will be critical for the EU to work collaboratively with other countries to build a global regulatory framework that addresses the ethical and technological challenges of AI.

4.2 Flexibility and Adaptability

The Regulation is also designed to be flexible and adaptable, recognizing the dynamic and evolving nature of AI. For this reason, it includes provisions that allow for periodic reviews and updates to address new technological and societal challenges. In addition, the regulation allows for codes of conduct and best practices to be developed in collaboration with industry, which facilitates the adaptation of the rules to different contexts and sectors (López-Tarruella 2023).

Artificial intelligence (AI) is a rapidly growing technology with an impact on multiple sectors, from the economy to the field of human rights. Faced with the challenges posed by this emerging technology, the European Union has taken a leadership position in regulating AI through its Artificial Intelligence Regulation (known as the EU AI Law). This regulation aims to mitigate the risks associated with AI systems, promote transparency, and protect the fundamental rights of European citizens. However, one of the most important aspects of this regulation is its flexible and adaptive design, which seeks to balance technological innovation with regulation and allow businesses to effectively adapt to its requirements (Commission 2021).

- Flexibility in risk categorization: One of the features that make this regulation adaptable is its risk-based approach. Unlike more rigid regulations, the IA Regulation classifies IA systems into different risk categories: unacceptable, high, limited, and minimal. AI systems classified as unacceptable risk, such as those used in subliminal manipulation techniques or social scoring systems, are banned altogether. On the other hand, high-risk systems, such as those used in critical sectors like health, education, or recruitment, must meet stricter requirements for transparency, human oversight, and safety (Commission 2021).

This categorization of risks allows for more efficient and targeted regulation, where regulatory resources are primarily targeted at those AI systems that pose the greatest risks to people's rights and safety. This flexible approach allows companies to tailor their systems and processes according to the level of risk, which facilitates compliance and reduces the regulatory burden on low-risk systems, thereby encouraging innovation in areas where risks are minimal (Veale and Zuiderveen Borgesius 2021).

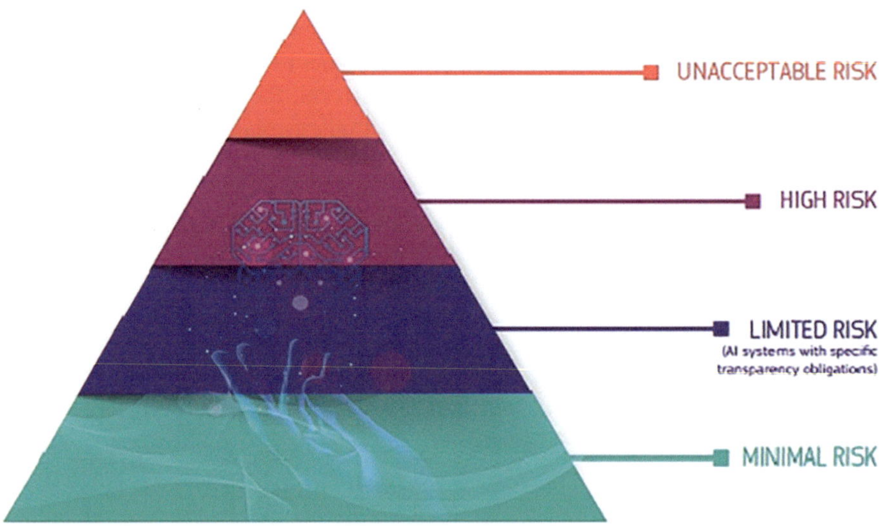

Source European Commission

- Adaptability in transparency and explainability requirements: The AI Regulation also promotes transparency and explainability in AI systems, especially in those considered high risk. However, the regulation recognizes that the levels of transparency and explainability required may vary according to the context and the specific use of each system. In this regard, the regulation allows for a certain degree of adaptability in the implementation of these requirements. For example, an AI system used in the financial sector may require a higher level of explainability than a recommendation system on an entertainment platform, given the potential impact on users and society (Wulf and Seizov 2020).

Furthermore, the regulation allows companies to implement transparency measures according to the technical capabilities of each system. This means that developers are not obliged to implement transparency requirements uniformly but can adapt their mechanisms to comply with the standards reasonably and effectively, depending on the technical characteristics and context of the use of the AI system. This flexibility encourages greater adoption of transparency measures without imposing excessive restrictions that could inhibit innovation (Busuioc et al. 2023).

- Exemptions and adaptability in compliance: Another aspect reflecting the flexibility of the IA Regulation is the possibility to apply exemptions in certain specific cases. The regulation provides exemptions for situations where compliance with the requirements may be too burdensome or inappropriate, such as in the case of early-stage research and development projects. These exemptions allow AI research to continue without regulatory barriers that could restrict experimentation and progress in innovative technologies. However, these exemptions are conditional on the projects meeting certain guidelines and not posing significant risks to citizens' rights (De Bruyne and Vanleenhove 2021).

This adaptability is also seen in the regulation's willingness to adjust its requirements in line with technological developments and new knowledge about AI risks. The European Commission can update and modify the regulation's compliance standards as technology evolves, allowing the regulation to remain relevant and effective in an ever-changing technological environment.

- Public–private cooperation: The flexibility of the IA Regulation is not only reflected in its internal structure but also in the collaborative approach it encourages between public and private sector actors. The European Commission has stressed the importance of working in partnership with AI companies, academia, and other stakeholders to implement the Regulation effectively. This collaborative approach allows stakeholders to contribute their knowledge and experience, which facilitates the adaptation of regulatory requirements to the needs and realities of the sector (Commission 2021).

In addition, forums and working groups have been established to guide companies on the implementation of the regulation, which helps to ensure that the regulation is applied consistently across sectors and that companies have the necessary support to adapt to the requirements. This collaboration is critical to effective regulatory implementation, as it allows for an exchange of knowledge and practices that benefit both regulators and AI developers (Commission 2024).

- The role of flexibility in innovation and competitiveness: One of the objectives of the AI Regulation is to encourage innovation in the development of safe and reliable AI technologies. Flexibility in the application of this regulation allows companies to innovate without fear of facing excessive regulatory burdens, which is especially important for startups and small businesses that might be constrained by strict regulation. By allowing for an adaptation of requirements depending on the nature of the AI system, the regulation promotes an environment where innovation can flourish within a framework of safety and accountability (Veale and Zuiderveen Borgesius 2021).

Moreover, this flexibility also positions the EU as a competitive space in the field of AI, where companies can develop advanced technologies while complying with high ethical and safety standards. Through this flexible regulation, the EU seeks to attract technology companies that are willing to work in a regulated but adaptive environment, which could strengthen the region's competitiveness in the global AI arena (Veale and Zuiderveen Borgesius 2021).

Finally, a regulatory framework is essential to build public trust in AI. Lack of regulation can generate skepticism and fear among users, who may fear the potential risks of the technology, such as loss of privacy, discrimination, or impact on employment. Regulation, by setting clear rules and protecting users' rights, helps reduce these fears and facilitates the safe adoption of AI in society (Busuioc et al. 2023).

Public trust is fundamental to the development of AI, as its widespread adoption depends to a large extent on users' perception of the technology as safe and beneficial. The EU AI Act, by requiring transparency, safety, and oversight in AI systems, contributes to strengthening this trust and enables the technology to be more safely and effectively integrated into society (Veale and Zuiderveen Borgesius 2021). Regulation of AI therefore not only protects users but also drives the growth of the technology by making it more acceptable and trustworthy.

4.3 Definition and Classification of Artificial Intelligence Systems

One of the most crucial and hotly debated aspects of the EU Artificial Intelligence Regulation is the definition and classification of AI systems. These foundational concepts are essential to the Regulation's implementation, as they determine which systems fall under its scope and the specific obligations imposed on them. The clarity and comprehensiveness of these definitions are vital for ensuring that the Regulation remains effective and relevant as AI technology advances.

The Regulation provides a broad yet adaptable definition of "artificial intelligence," acknowledging the diversity and complexity of technologies within this field. According to the Regulation, an AI system is defined as software that, using techniques such as machine learning, logic, and symbolic reasoning, can produce outcomes such as predictions, recommendations, or decisions that influence the environment with which it interacts (European Commission 2021). This definition intentionally covers a wide range of technologies, from advanced deep learning algorithms to traditional rule-based systems, allowing the Regulation to remain inclusive of various AI methodologies and applications.

The breadth of this definition serves to future-proof the Regulation, allowing it to remain applicable to both current and emerging AI technologies and preventing it from becoming obsolete as the field of AI continues to evolve rapidly. By being flexible, the definition enables the Regulation to encompass future technologies not yet fully developed or widely understood but that may have substantial societal or ethical implications. This approach reflects the EU's commitment to proactively address potential risks associated with AI (Floridi 2021).

Moreover, the Regulation's broad classification allows it to differentiate AI systems based on their risk level, rather than limiting the scope solely to specific applications or methodologies. This flexibility is crucial, as it enables regulators to focus on the impact and potential risks of AI applications, rather than rigidly defining AI based on current technology alone. Such an approach allows the Regulation to balance regulatory oversight with the flexibility needed for technological growth and innovation, ensuring that it can adapt to the changing landscape of AI (Commission 2021).

In sum, the EU's flexible and inclusive approach to defining and classifying AI within its regulatory framework is designed to protect fundamental rights while fostering innovation. By encompassing both established and emerging AI technologies, the Regulation sets a comprehensive foundation that can evolve in line with advancements in the field, supporting the EU's goals of responsible and forward-looking AI governance.

Despite its ambitious goals, the EU AI Regulation faces several challenges. One of the primary concerns is the potential difficulty of enforcing the regulation consistently across all EU member states. The Regulation requires each member state to designate competent authorities for monitoring and enforcing AI compliance, but variations in enforcement capacity and regulatory resources may lead to inconsistencies. Differences in budget allocations, regulatory expertise, and administrative support among member states could result in uneven application of the Regulation's standards, potentially undermining its objectives of harmonization and uniform protection across the EU (Veale and Zuiderveen Borgesius 2021). Furthermore, member states with more limited resources may struggle to keep pace with the Regulation's evolving requirements, creating disparities in compliance that could complicate cross-border collaboration and reduce overall regulatory effectiveness.

Additionally, businesses, particularly small and medium-sized enterprises (SMEs), may encounter challenges in meeting the Regulation's stringent technical requirements. These requirements cover areas such as documentation, explainability, and transparency—essential elements for ensuring accountability in high-risk AI applications. However, these standards may demand significant investments in technology and compliance resources. For SMEs, these operational costs could be prohibitive, as they often lack the same access to capital and expertise as larger corporations. The need for in-depth documentation and transparency may further strain resources, requiring SMEs to hire specialized personnel or invest in external compliance support, ultimately impacting their competitiveness (Parliament 2023). This burden could potentially create an uneven playing field, where only larger companies can afford to fully comply, potentially limiting innovation among smaller enterprises.

Another challenge lies in balancing innovation with regulation. While the EU has made efforts to create a flexible framework that supports growth, some stakeholders argue that the strict regulations on high-risk applications could inhibit AI innovation within Europe. Critics caution that overly restrictive compliance requirements may deter experimentation and development in areas such as healthcare, finance, and transportation—sectors where AI holds considerable promise. This potential stifling of innovation could lead European companies to lag behind competitors in regions with more lenient regulatory environments, such as the United States or certain Asian countries, where companies may have greater freedom to test and deploy new AI technologies (Floridi 2021). This competitive disadvantage might result in a "brain drain" of AI talent and a shift in investment toward markets with fewer restrictions, which could weaken Europe's leadership in AI innovation.

To address these concerns, the EU has committed to periodic reviews of the Regulation, allowing for adjustments in response to technological advances and market developments. By adopting an iterative approach to regulation, the EU aims to create a dynamic framework that can evolve alongside the rapidly changing landscape of AI technology. These periodic reviews offer an opportunity for the EU to assess the Regulation's impact on innovation and competition, making necessary modifications to ensure that it remains supportive of economic growth while maintaining rigorous ethical and safety standards (Commission 2021). This adaptability is critical for maintaining the Regulation's relevance and effectiveness, as it allows policymakers to respond to new insights, emerging technologies, and feedback from industry stakeholders.

In conclusion, the EU Artificial Intelligence Regulation is a comprehensive and forward-looking framework that aims to protect citizens' rights, ensure ethical AI practices, and support innovation. Through its risk-based approach, transparency obligations, and extraterritorial application, the Regulation sets a high standard for responsible AI use, establishing the EU as a leader in digital governance and ethical AI development. Although challenges remain in implementing and enforcing the Regulation, particularly in terms of consistent application across member states and the balance between regulation and innovation, the EU's commitment to periodic reviews and adaptable policy structures demonstrates its dedication to fostering safe, transparent, and accountable AI (Smuha 2021). The Regulation's potential to shape global AI practices underscores its significance as a model for future regulatory efforts worldwide, positioning the EU at the forefront of the global discourse on responsible AI governance.

5 Conclusions

The EU's Artificial Intelligence Regulation represents a significant milestone in global efforts to regulate artificial intelligence within a framework of accountability and ethics. Its risk-based approach, strict compliance requirements, and promotion of harmonized standards aim not only to protect citizens but also to foster competitiveness and responsible innovation in the development of AI technologies. Through this regulatory framework, the EU protects fundamental rights, addressing both present and future challenges, while setting an influential precedent for how the world can navigate the complexities of AI.

The EU AI Regulation is one of the most advanced regulatory initiatives in the field of artificial intelligence, notable for its flexible and adaptive approach. By allowing companies to meet regulatory requirements in a proportionate and contextually appropriate manner, the Regulation facilitates both innovation and safety in the deployment of AI systems. The risk classification system, tailored transparency requirements, and specific exemptions create a nuanced regulatory structure that adapts to the rapidly evolving technological landscape. Furthermore, by fostering public–private partnerships, the Regulation enhances effective implementation, ensuring that security and fundamental rights are upheld without imposing undue constraints on technological progress.

This flexible and forward-thinking regulatory framework positions the EU as a global leader in the ethical and safe governance of AI. The EU's approach to balancing citizen rights protection with the promotion of innovation reflects a commitment to creating an environment where AI technologies can thrive responsibly. This combination of adaptability and rigor not only benefits AI companies by providing a predictable regulatory landscape but also instills confidence in European citizens that AI technologies are being developed and used responsibly and ethically. Through this leadership, the EU sets a benchmark for international AI governance, encouraging other regions to adopt similar standards that prioritize transparency, accountability, and public trust.

Additionally, this regulation intersects with broader goals in information security, cybersecurity culture, and data protection. The EU's Cybersecurity Strategy highlights the interconnectedness of these domains, emphasizing the importance of building a strong security culture across member states. As cyber threats evolve, maintaining a cohesive approach to cybersecurity and data protection remains a priority for the EU, reinforcing the need for public awareness and resilient systems.

In alignment with these priorities, we recommend further steps to promote cyber hygiene and cybersecurity literacy across the EU:

1. **Design Incentive Programs for National Education Authorities**: Member states' national education authorities should be incentivized to integrate cybersecurity into all levels of public education, starting from early childhood education. Providing structured cyber literacy from a young age establishes a foundational understanding of cybersecurity best practices.

2. **Promote Lifelong Learning Programs for Cybersecurity**: Cybersecurity education should extend beyond the traditional school years. Lifelong learning initiatives should target citizens outside of the workforce and those over 50 years of age, who may be particularly vulnerable to cyber threats due to limited exposure to digital technologies.
3. **Develop Free Online and Onsite Short Courses**: Accessible, free cybersecurity courses should be offered to the general public, covering essential topics such as safe online practices, recognizing phishing attempts, and protecting personal information. Additionally, making such courses mandatory within public education institutions would further enhance the EU's cybersecurity culture.
4. **Encourage Entrepreneurship and Research in Cybersecurity**: At the EU level, programs should support cybersecurity-focused entrepreneurship and research, targeting a broad range of organizations but with special attention to small and medium-sized enterprises (SMEs), which may lack the resources for comprehensive cybersecurity measures. Such programs can drive innovation and help SMEs strengthen their cybersecurity infrastructure.

These recommendations address the human element, which should remain at the core of EU policies and regulations in cybersecurity. By investing in education, awareness, and innovation, the EU can cultivate a cybersecurity-conscious society equipped to face future challenges. This focus on human factors complements the technical standards and legal frameworks established in the AI Regulation, contributing to a comprehensive strategy for responsible and secure digital development in Europe.

In summary, the EU AI Regulation represents a transformative framework that not only protects citizen rights but also promotes an inclusive and secure environment for AI innovation. With the addition of cybersecurity initiatives and education-focused policies, the EU is well-positioned to lead in both digital ethics and resilience, setting an example for a globally responsible approach to the future of artificial intelligence and digital security.

References

Benini, C. (2022). A Private International Law Perspective on the Extraterritorial Reach of EU Law: The "Docile" Attitude of EU Conflict of Law Rules. *Centre for the Law of EU External Relations Papers*, 25–36. https://hdl.handle.net/10807/213727*]*

Birchfield, V. L. (2024). From roadmap to regulation: will there be a transatlantic approach to governing artificial intelligence? *Journal of European Integration*, *46*(7), 1053–1071. https://doi.org/10.1080/07036337.2024.2407571

Busuioc, M., Curtin, D., & Almada, M. (2023). Reclaiming transparency: contesting the logics of secrecy within the AI Act. *European Law Open*, *2*(1), 79–105. https://doi.org/10.1017/elo.2022.47. https://www.cambridge.org/core/journals/european-law-open/article/reclaiming-transparency-contesting-the-logics-of-secrecy-within-the-ai-act/01B90DB4D042204EED7C4EEF6EEBE7EA

Butcher, J. G., & Beridze, I. (2019). What is the state of artificial intelligence governance globally? *The RUSI Journal*, ***164*****(5-6), 88-96.** https://doi.org/10.1080/03071847.2019.1694260

Calo, R. (2017). *Artificial Intelligence Policy: A Primer and Roadmap.* University of California Law Review, 57, 273–312. https://lawreview.sf.ucdavis.edu/sites/g/files/dgvnsk15026/files/media/documents/51-2_Calo.pdf

Czerniawski, M. (2024). T*owards the effective extraterritorial enforcement of the AI Act*. https://www.researchgate.net/publication/384292786_Towards_the_effective_extraterritorial_enforcement_of_the_AI_Act

De Bruyne, J., & Vanleenhove, C. (Eds.). (2021). Artificial intelligence and the law. Intersentia. https://doi.org/10.1017/9781839701047

Dignum, V. (2023, January 1). Responsible Artificial Intelligence: Recommendations and Lessons Learned. Springer International Publishing, 195–214. https://doi.org/10.1007/978-3-031-08215-3_9

European Commission. (2018). *Artificial intelligence for Europe* (COM(2018) 237 final). European Union. https://eur-lex.europa.eu/legal-content/EN/ALL/?uri=CELEX:52018DC0237

European Commission. (2020). *White Paper on Artificial Intelligence - A European approach to excellence and trust*. https://ec.europa.eu/info/publications/white-paper-artificial-intelligence-european-approach-excellence-and-trust_en

European Commission. (2021). *Artificial Intelligence Act: The EU plan for trustworthy AI*. https://ec.europa.eu/commission/presscorner/detail/en/IP_21_1682

European Commission. (2021). *Proposal for a Regulation laying down harmonised rules on Artificial Intelligence (Artificial Intelligence Act).* https://eur-lex.europa.eu/legal-content/EN/TXT/?uri=CELEX:52021PC0206

European Parliament. (2023). *The EU Artificial Intelligence Act and its implications for SMEs and innovation in Europe.* EU Publications

European Commission. (2024). Meet the Chairs leading the development of the first General-Purpose AI Code of Practice. https://digital-strategy.ec.europa.eu/en/news/meet-chairs-leading-development-first-general-purpose-ai-code-practice

Floridi, L. (2021). The European legislation on AI: A brief analysis of its philosophical approach. Philosophy & Technology, 34(2), 215–222. https://doi.org/10.1007/s13347-021-00460-9

Goodman, B., & Flaxman, S. (2017). European Union regulations on algorithmic decision-making and a "right to explanation". AI Magazine, 38(3), 50-57. https://doi.org/10.1609/aimag.v38i3.2741

Hamon, R., Junklewitz, H., & Sanchez Martin, J. I. (2020). Robustness and explainability of artificial intelligence: From technical to policy solutions. *Publications Office of the European* Union. https://doi.org/10.2760/57493

Hauer, M P., Krafft, T D., Sesing-Wagenpfeil, A., & Zweig, P K. (2023, January 1). Quantitative study about the estimated impact of the AI Act. Cornell University. https://doi.org/10.48550/arxiv.2304.06503

Huang, J Y., Gupta, A., & Youn, M. (2021, March 23). Survey of EU ethical guidelines for commercial AI: case studies in financial services. Springer Nature, 1(4), 569–577. https://doi.org/10.1007/s43681-021-00048-1

López-Tarruella Martínez A. (2023). El futuro Reglamento de Inteligencia Artificial y las relaciones con terceros Estados. Revista electrónica de estudios internacionales (REEI), ISSN-e 1697–5197, Nº. 45. http://www.reei.org/index.php/revista/num45/articulos/futuro-reglamento-inteligencia-artificial-relaciones-con-terceros-estados

Smuha, N. A. (2021). From a 'race to AI' to a 'race to AI regulation': Regulatory competition for artificial intelligence. *Law, Innovation & Technology, 13*(1), 57-84. https://doi.org/10.1080/17579961.2021.1898300

Tartaro, A., Smith, A L., & Shaw, P. (2023, January 1). Assessing the impact of regulations and standards on innovation in the field of AI. Cornell University. https://doi.org/10.48550/arxiv.2302.04110

Veale, M., & Zuiderveen Borgesius, F. J. (2021). Demystifying the draft EU Artificial Intelligence Act: Analysing the good, the bad, and the unclear elements of the proposed approach. *Computer Law & Security Review, 39*, 105531. https://doi.org/10.48550/arXiv.2107.03721

Wong, A. (2021). Ethics and Regulation of Artificial Intelligence. In: Mercier-Laurent, E., Kayalica, M.Ö., Owoc, M.L. (eds) Artificial Intelligence for Knowledge Management. AI4KM 2021. IFIP Advances in Information and Communication Technology, vol 614. Springer, Cham. https://doi.org/10.1007/978-3-030-80847-1_1

Wulf, A. J., & Seizov, O. (2020). Artificial intelligence and transparency: A blueprint for improving the regulation of AI applications in the EU. *European Business Law Review, 31*(4), 611–640. https://doi.org/10.54648/eulr2020024

International and National Perspective of the Digitalization of the Spanish Justice Administration in the Criminal Process: Impact on Fundamental Rights

Monica Pucci-Rey

Abstract This research examines the impact of digitalization in the criminal justice system, highlighting the advances and challenges arising from the implementation of new technologies in judicial procedures. Through an analysis of Organic Law 1/2025, Royal Decree 6/2023, and various key rulings from the Supreme Court and the Constitutional Court, it addresses topics such as the right to due process, personal data protection, and the technological gaps that affect equality of arms. The article concludes with proposals for improvement, including legal harmonization and technological training for legal practitioners to ensure justice that is more accessible, efficient, and respectful of fundamental rights.

1 Introduction

The Administration of Justice constitutes one of the three powers of the State, recognized in the Spanish Constitution of 1978 as a guarantor of fundamental rights and freedoms. Within it, the criminal jurisdiction plays an essential role in ensuring the protection of public order and the prosecution of crimes, guaranteeing the principle of legality and the right to a process with all guarantees. Its independence and autonomy are essential for judicial decisions to be objective, impartial, and respectful of the principles of the rule of law.

The digital transformation of the Administration of Justice emerges as an unavoidable imperative to adapt this essential pillar to the challenges and opportunities of the contemporary era. Currently, society is undergoing a profound transformation driven by the technological revolution. The omnipresence of digital technology has permeated all aspects of daily life, from communication to business management. In this context, the digital transformation of the Administration of Justice has ceased to be an option to become an imperative necessity, especially in the criminal sphere,

M. Pucci-Rey (✉)
Rey Juan Carlos University (URJC), Madrid, Spain
e-mail: monica.pucci@urjc.es

L. A. Garcia-Segura (ed.), *Collaborative Insights on Technology and Legal Frameworks*, Advanced Sciences and Technologies for Security Applications,
https://doi.org/10.1007/978-3-032-27714-5_4

where procedural agility and efficiency are fundamental to guarantee effective judicial protection. Contemporary society is marked by technological advancement, and justice cannot remain outside this structural change.

Electronic Judicial Administration refers to the use of information and communication technologies (ICT) in the judicial field to manage cases, files, communications, and procedures electronically. This approach encompasses a variety of practices, from the electronic submission of documents to the holding of online hearings and trials. The digitalization of judicial processes aims primarily to improve the efficiency, accessibility, and transparency of the justice system, adapting it to the demands of an increasingly connected society.

The use of information and communication technologies in the judicial field is known as Digital Justice Administration, E-Justice, or Electronic Justice., as Medina Zepeda (2022, p. 177) has pointed out:

> *Digital justice, electronic justice, virtual justice, cyberjustice (from the French cyberjustice) or, adapted from Anglo-Saxon law, E-justice, are expressions of increasingly widespread use to refer to the application of digital technology in the administration and imparting of justice. However, E-justice is often confused with other related concepts (for example, with online trials), which is why its conceptualization is essential.*

It is evident from what has already been stated that the concept of Electronic Judicial Administration is broad, as it does not focus on a single aspect of procedural practice. González Campo (2013, p. 36) has defined it as:

> *That set of organizational, legal, and technical—now electronic—means or resources that, through an ordered and concrete implementation in judicial processing, provide it with greater efficiency and effectiveness, and are placed at the service of the jurisdictional function so that it has better means to, ultimately, respond to the exercise by the litigant of the fundamental right to effective judicial protection recognized in Article 24 of the Spanish Constitution.*

The key to this definition lies in analyzing the set of organizational, legal, and technical means and resources. In the organizational sphere, Electronic Judicial Administration seeks to establish a coherent order that allows for more efficient case management. For example, the implementation of electronic case management systems, such as those currently offered by the HORUS Electronic Portal, facilitates the organization and tracking of judicial files more effectively. The creation of electronic portals for access to judicial information not only streamlines the consultation of relevant data but also provides legal professionals and citizens with a clearer and more accessible view of ongoing judicial procedures.

In the legal field, Electronic Judicial Administration acts as a catalyst for the adaptation and improvement of regulatory frameworks. The incorporation of electronic signatures as a valid tool for the submission of judicial documents is a tangible example of how legal resources are adjusted to accommodate digitalization. In the technical field, Electronic Judicial Administration relies on cutting-edge tools and technologies. Videoconferencing platforms for holding virtual hearings are a concrete example of how technical resources enrich the jurisdictional function. This approach not only streamlines processes but also reduces the need for physical travel, optimizing resources and guaranteeing more accessible justice.

In Spain, the modernization of justice has been accompanied by the promulgation of various norms aimed at implementing technological tools in judicial procedures. The most recent is Organic Law 1/2025, of March 1, which introduces important modifications to the Organic Law of the Judiciary (LOPJ) and to procedural laws to consolidate the digitalization of criminal justice. This reform affects key aspects such as the management of digital files, the incorporation of telematic tools in criminal proceedings, and the protection of personal data in the judicial sphere. Added to this is Royal Decree-Law 6/2023, of December 19, which, in its Book I, develops digital and procedural efficiency measures in the public service of justice. Díaz (2023), one of the many voices that have analyzed this relatively recent norm in the digital medium El Economista, highlights that this norm:

> *Brings relevant changes such as the obligation to prioritize the holding of telematic trials instead of face-to-face ones (...) and changes the way citizens and companies will have to interact with the Administration, facilitating the digital option for the former, but obliging the latter to carry out any procedure through this channel.*

The preamble of Royal Decree-Law 6/2023 highlights that the digitalization of the judicial system seeks to optimize procedural times, reduce bureaucratic burden, and improve citizens' access to justice. However, the practical implementation of these measures has revealed significant difficulties, which we will particularly study throughout this work in the criminal sphere.

In short, we can conclude that Electronic Judicial Administration is presented as a complex framework of means that converges to enhance justice, directly influencing judicial practice: from digital platforms that facilitate remote work from a computer like LexNET, to electronic communications between different administrations that shorten the time of transmission and reception of information, streamlining judicial procedures and avoiding unnecessary delays that can ultimately cause defenselessness. Electronic Judicial Administration is not just a technological evolution; it is an integral evolution in the way we conceive and experience the administration of justice.

The present work is framed in a legal-theoretical and practical approach, combining normative analysis with the study of its real application in criminal jurisdiction. The methodology used is based on the following main axes: normative and doctrinal analysis; study of jurisprudence; comparative law; critical analysis of judicial practice; proposals for improvement, since, based on the normative and practical analysis, proposals will be formulated to improve the digitalization of criminal justice in Spain, addressing issues such as legislative harmonization, interoperability of technological systems, and the training of legal operators.

2 International and National Regulations on the Digitalization of Criminal Proceedings

The digitalization of the Administration of Justice has been a process driven by the national and international regulatory framework with the aim of modernizing the judicial system and making it more efficient. This process, although led by the State under Article 149.1.5a of the Spanish Constitution (EC), involves multiple actors, such as the General Council of the Judiciary, the Ministry of Justice, the Public Prosecutor's Office, and the Autonomous Communities with jurisdiction over justice.

Electronic Judicial Administration in Spain has undergone significant regulatory evolution, with the convergence of national and European legislation seeking to modernize and streamline the judicial system. This has ranged from the first European guidelines to the most recent national regulations.

The European Union has been a key driver in the creation of e-court administration in its member states, including Spain. Directive 1999/93/EC of the European Parliament and of the Council of 13 December 1999 on a Community framework for electronic signatures marked a milestone by establishing a legal framework for the use of recognized electronic signatures throughout the EU. Furthermore, Directive 2000/31/EC of the European Parliament and of the Council of 8 June 2000 on certain legal aspects of information society services further consolidated legal support for the use of electronic documents in court proceedings. Numerous European regulations govern this matter, including the following:

- Regulation (EU) 910/2014 eIDAS (Electronic Identification, Authentication and Trust Services) establishes the legal framework for electronic identification and trust services in electronic transactions within the European Union. In the criminal field, this regulation is essential because it regulates the validity of qualified electronic signatures, allowing their recognition in all Member States and facilitating the presentation of digital documents and evidence in criminal proceedings. It promotes the use of digital identity in the identification of parties in criminal proceedings, facilitating secure access to digital files.
- Directive 2016/343/EU: (Presumption of innocence and right to be present at the trial) This directive strengthens the fundamental right to a fair trial and regulates key aspects of the digitalization of criminal proceedings. In particular, it allows the use of electronic means to ensure the appearance of the accused at trial without undermining their right to a defense and reinforces the need for any electronic communication between the Administration of Justice and the accused to respect their right to be informed of the charges and to have adequate time and resources to prepare their defense.
- Regulation (EU) 2018/1727 (Eurojust): This regulation regulates the functioning of Eurojust, the EU agency for criminal judicial cooperation, and its ability to exchange information digitally in cross-border criminal investigations and prosecutions. Its key aspects include implementing an electronic case management system between prosecutors and judges in different Member States to improve

judicial cooperation, obliging Member States to provide remote and electronic access to court files of interest to Eurojust, while respecting data protection safeguards, and allowing the electronic processing of European arrest warrants (EAWs) and European investigation orders (EIOs), reducing response times in criminal cooperation.
- Directive 2013/48/EU: Guarantees access to a lawyer in criminal proceedings, including the possibility of telematic assistance. This regulation requires Member States to adopt electronic systems for notification of rights and legal assistance, ensuring the confidentiality and accessibility of information.
- Regulation (EU) 2020/1783: While this regulation focuses on judicial cooperation in civil matters, it also has important implications for criminal jurisdiction, as it introduces the possibility of conducting remote evidence and witness statements via videoconference in criminal proceedings.
- Framework Decision 2008/978/JHA: Regulates the European Electronic Evidence Order.
- EU e-Justice Programme: Seeks the digitalization of justice throughout the Union, promoting tools such as the e-Justice Portal for the exchange of judicial documentation.

Within the international framework, we can also find numerous Conventions and Pacts that must govern our positive law:

- European Convention on Human Rights (ECHR): Article 6 establishes the right to a fair trial, which implies that judicial digitalization must guarantee equality of arms and immediacy. The European Court of Human Rights has stated that "participation in proceedings by videoconference is not, in itself, incompatible with the concept of a fair and public trial, but it is necessary to ensure that the person concerned can follow the proceedings and be heard without technical obstacles and can communicate effectively and confidentially with his or her lawyer" (ECHR, judgment of November 2, 2010, Sakhnovski v. Russia)
- Budapest Convention on Cybercrime (2001): Facilitates international cooperation in the obtaining of electronic evidence and in the investigation of computer-related crimes.
- European Convention on Mutual Assistance in Criminal Matters (1959): Its Second Additional Protocol of 2001 incorporates the use of technologies for remote testimony-taking.
- International Covenant on Civil and Political Rights (ICCPR): Protects the right to a fair trial in Article 14, which applies to digital proceedings.
- United Nations Convention against Corruption (UNCAC, 2003): Urges States to use electronic tools to improve efficiency and transparency in the criminal justice system.

At the national level, Law 34/2002, of July 11, on Information Society Services and Electronic Commerce (LSSI), played a crucial role in establishing the legal framework for the use of electronic means in the judicial sphere. This law provided the basis for the electronic filing of documents, telematic communications, and other aspects related to Electronic Judicial Administration. Subsequently, Royal Decree-Law 3/2010, of March 8, regulating the National Security Framework in the field of Electronic Administration, established guidelines to guarantee the security of systems and data in Public Administration, thus laying the foundations for trust in the use of electronic documents and digital signatures in the judicial sphere.

However, the main regulatory milestone in the creation of e-Judicial Administration in Spain lies in Law 18/2011, of July 5, regulating the use of information and communications technologies in the Administration of Justice (currently repealed by Royal Decree-Law 6/2023). This law established a comprehensive framework for the incorporation of information and communications technologies in the administration of justice, covering everything from the electronic filing of documents to the execution of procedural acts by electronic means. Law 18/2011 not only defined the bases for the use of electronic signatures but also made the electronic filing of documents and documents mandatory for certain procedural subjects. Furthermore, it introduced the concept of the electronic judicial file, enabling the electronic management of judicial proceedings and promoting greater efficiency in the processing of cases.

Law 41/2015, which amends the Criminal Procedure Law (LECrim), introduced a series of advances in the digitalization of criminal proceedings, allowing for greater efficiency and agility in the management of judicial proceedings. Its key reforms include the electronic filing of complaints and lawsuits and the electronic collection and transmission of evidence. This allowed individuals or legal entities to file complaints or lawsuits without having to physically appear in court, expediting the initiation of criminal proceedings and facilitating access to justice, especially in hard-to-reach territories. The implementation of this measure also contributed to greater efficiency in the management of the volume of judicial proceedings, allowing proceedings to begin without unnecessary delays. The reform also allowed for the electronic collection and transmission of evidence between law enforcement and judicial bodies. This facilitated the immediate and secure transmission of documents, reports, and evidence, reducing waiting times and human errors that could occur when managing documents in physical format. This allows law enforcement agencies and courts to work more coordinated and efficiently in the collection and analysis of digital evidence, which is especially relevant in the field of cybercrime or when evidence is obtained in digital formats.

Furthermore, Law 41/2015 regulated how these interventions in electronic communications should be carried out, ensuring their procedural validity and respecting fundamental rights, especially with regard to the protection of privacy and procedural guarantees of those involved. This reform became an important pillar in the context of rising digital crime, allowing judges to authorize electronic interventions in a more agile manner, respectful of rights, and in line with technological advances.[1]

Royal Decree 1065/2015 regulates the use of electronic communications in the Administration of Justice in Spain, facilitating the management and processing of

[1] See articles 265, 282 bis and 579 of the Criminal Code, respectively, as amended by Law 41/2015.

judicial proceedings through digital means. The Royal Decree establishes the mandatory use of the LexNET system for the electronic filing and notification of procedural documents. LexNET is the official platform that allows legal practitioners (lawyers, solicitors, prosecutors, judges, etc.) to carry out their procedures digitally, without the need to physically present documents in court. This system has facilitated efficient communication between the various actors in criminal proceedings, reducing processing times and minimizing the risk of loss or misplacement of important documentation. The regulation also requires lawyers, solicitors, and other legal practitioners to conduct all their judicial proceedings electronically. This measure has been especially relevant in the criminal field, where the processing of proceedings is often voluminous and urgent. The digitalization of judicial proceedings has allowed for greater efficiency in case management and facilitated faster access to information, allowing those involved in criminal proceedings to receive electronic notifications quickly and securely.[2]

Organic Law 7/2021, approved in 2021, is a fundamental step in the protection of personal data in the criminal field. This law adapts Spanish regulations to the requirements of Regulation (EU) 2016/679 of the European Parliament and of the Council on the protection of personal data (General Data Protection Regulation, GDPR) and regulates key aspects related to data management in criminal proceedings. It establishes specific regulations on how personal data should be managed in the criminal field, especially when it involves information obtained digitally, such as electronic communications records, banking information, or personal data linked to those under investigation. The law establishes that data processed in the criminal field must be processed legally, transparently, and in a manner that guarantees its security. It also regulates the right to access, rectify, erase, and object to personal data in criminal proceedings. The law also regulates access to criminal databases by judges, prosecutors, and other judicial actors through electronic means. Access is controlled, guaranteeing the confidentiality and protection of information and ensuring that only authorized legal practitioners can access this data. This allows for the streamlining of the collection of criminal records, arrest reports, or electronic records that may be relevant to the development of criminal investigations.[3]

Royal Decree-Law 6/2023, of December 19, and Organic Law 1/2025, of January 9,23 have marked key milestones in the digital transformation of criminal justice in Spain. These regulations have introduced profound changes to procedural legislation, driving the modernization of the judicial system through the digitalization of procedures, improved interoperability between judicial bodies, and the consolidation of advanced technological tools.

Royal Decree-Law 6/2023, which entered into force on March 20, 2024, establishes a cross-cutting regulatory framework affecting all jurisdictions, with a particular impact on criminal jurisdiction. This regulation reinforces the mandatory use of electronic means in procedural processing, expanding the scope of electronic court files and consolidating interoperability between administrations. Furthermore,

---

[2] Articles 6 to 9 of Royal Decree 1065/2015.

[3] See articles 4, 12 and 24 of Organic Law 7/2021.

it regulates remote and secure access to judicial proceedings for the parties involved, guaranteeing the right to defense in digital environments and strengthening data protection in the management of criminal information. One of the most notable aspects is the update of the regime for the intervention of electronic communications, introducing new procedural guarantees and strengthening judicial oversight in its application.

For its part, Organic Law 1/2025, which entered into force in January 2025, complements and strengthens the provisions of Royal Decree-Law 6/2023, establishing essential principles to guarantee legal certainty in the use of technologies in the criminal field. Its regulation focuses on the digitalization of procedural proceedings and the adoption of tools such as videoconferencing in criminal trials, electronic access to court records, and the use of advanced case management systems. Furthermore, it introduces improvements to the regulations on the processing of personal data in criminal proceedings, ensuring compatibility with Regulation (EU) 2016/679 (GDPR) and Organic Law 7/2021 on Data Protection in Criminal Matters. The current regulations on statements by videoconference arose from the COVID-19 pandemic, which required a necessary adaptation of procedural proceedings to the current reality. Unprecedented measures were adopted, such as the suspension of procedural deadlines, with some exceptions in the criminal jurisdiction, the holding of acts, appearances and hearings preferably electronically, etc. (Rayón, 2022, p. 289).

Given the impact of these reforms, a detailed analysis of them is addressed throughout this paper, examining their practical implications for the processing of criminal proceedings and their contribution to the efficiency and accessibility of the Administration of Justice.

3 Impact of the Digitalization of the Criminal Process on Fundamental Rights

The digitalization of judicial processes has significantly transformed the way criminal proceedings are managed, bringing clear benefits in terms of efficiency, agility, and accessibility. However, this modernization of the judicial system poses a set of challenges related to the fundamental rights of defendants, which are enshrined in the Constitution and international human rights treaties. These rights are the foundation upon which a fair trial is built and must be respected at all times, even when procedures are adapted to new technologies.[4]

Among the most relevant rights of defendants in the criminal context are the right to due process, which guarantees that all parties receive equal treatment during the trial; the right to defense, which includes the right to be informed of the charges and to have the necessary means to prepare a defense; and the right to a fair and public trial,

[4] See Article 24 of the Spanish Constitution, Article 6 of the European Convention on Human Rights (ECHR), Articles 47 and 48 of the Charter of Fundamental Rights of the European Union (CFREU), or Article 14 of the International Covenant on Civil and Political Rights (ICCPR).

which ensures that the accused can face justice in a transparent manner. Furthermore, immediacy and equality of arms are key principles that must be protected at all times, allowing the accused the opportunity to interact directly with witnesses, evidence, and the parties involved in the trial.

The incorporation of digital platforms and technologies such as video conferencing, electronic case management, and the use of digital evidence introduces new dynamics that can have a significant impact on the effectiveness and protection of these rights. While these tools can improve the efficiency and accessibility of the judicial system, they can also jeopardize transparency, inclusion, and equity if not properly managed.

This section addresses the impact of digitalization on the fundamental rights of defendants, analyzing how technological advances may affect their right to a fair trial, with particular attention to issues of procedural immediacy, access to information, and personal data protection. Throughout this analysis, both the benefits and risks associated with the implementation of digitalization in the criminal justice system will be examined, considering the necessary safeguards to protect fundamental rights in a context of increasing use of digital technologies.

3.1 The Criminal Process in a Digital Environment

The digital transformation has impacted every phase of criminal proceedings, from the preliminary investigation to the trial. Below, we analyze the impact of these reforms on each phase.

3.1.1 Instruction Phase

The purpose of the investigation phase is to investigate the crime and identify those responsible. Digitalization has affected this phase in several ways.

First, as we have already seen in detail, electronic access to court files is provided for in Article 234 of the LOPJ (Journal of Criminal Procedure), which establishes the right of the parties to access procedural information digitally. However, in the criminal field, access must guarantee the protection of the parties' rights, especially in cases of confidentiality of proceedings (Article 302 of the LECrim (Criminal Procedure Act). Furthermore, digital evidence has gained prominence in criminal proceedings, generating the need to establish standards of authenticity and integrity, as we have already analyzed. In STS 300/2015, the Supreme Court analyzed the requirements for the validity of digital evidence, concluding that authentication mechanisms and a chain of custody must be guaranteed for its admissibility.

The purpose of the investigative phase of a criminal case is to gather evidence and determine whether there is sufficient evidence to open a trial. In this context, checking criminal records through electronic records is a fundamental tool for assessing the recidivism of the person under investigation and adopting appropriate precautionary measures, especially in cases of gender-based violence, where the protection of the victim is a priority.

Access to criminal records is regulated by Article 136 of the Criminal Code, which establishes the time limits for expungement of records, and by Articles 610 and 611 of the Criminal Procedure Law (LECrim), which empower the investigating judge to request information from official registries during the investigation. This data is accessed through the System of Administrative Records to Support the Administration of Justice (SIRAJ), which allows interconnection between courts and other official bodies. Digitization has facilitated the consultation of criminal records during the investigation phase, allowing judges to make more informed decisions and effectively protect victims of gender-based violence. However, its use must comply with criteria of legality and proportionality, avoiding abuse and ensuring a balance between public safety and the fundamental rights of the person under investigation. Technological developments in this area, combined with appropriate regulation, will contribute to improving the effectiveness of the justice system without undermining procedural guarantees.

In matters of gender-based violence, access to these records is strengthened by Organic Law 1/2004 on Comprehensive Protection Measures against Gender-Based Violence, which imposes the obligation to review the criminal record of the person under investigation before adopting any precautionary measure. Furthermore, Directive 2012/29/EU on the rights of victims in criminal proceedings establishes the need for judicial authorities to have complete information on the aggressor's criminal record to ensure effective protection.

In proceedings for gender-based violence, reviewing criminal records is crucial for assessing the risk to the victim. The existence of prior convictions or current precautionary measures may justify the adoption of protective measures such as restraining orders or pre-trial detention, in accordance with Article 544 ter of the LECrim (Criminal Criminal Procedure Act).

The use of electronic records is not limited to the national level. In cases where the suspect may have committed crimes in other European Union Member States, Spanish authorities can access their criminal record through the European Criminal Records Information System (ECRIS),[5] regulated by Council Framework Decision 2009/315/JHA and Regulation (EU) 2019/816. Transnational access to this information is essential in gender-based violence crimes, as it allows for the detection of criminal records in other countries that may not be recorded in Spain.

Furthermore, the reform of the LECrim introduced measures that allow the use of digital means in criminal investigations, especially with regard to the interception of communications and remote access to electronic devices. These measures are regulated in:

- Article 588 bis LECrim: Permits the interception of electronic communications under judicial control.

[5] See https://www.mjusticia.gob.es/es/servicio-justicia/proyectos-transformacion/Foro-Transformacion/Dimensi%C3%B3n-internacional/ECRIS.

- Article 588 septies LECrim: Regulates remote access to computer systems as an investigative measure, establishing that it may only be authorized by a reasoned court ruling.

This last article is especially relevant, as it imposes the need for strict judicial oversight in the use of remote access technologies. Case law has indicated that the use of these tools must not violate the right to privacy or the principle of proportionality. Article 588 bis of the LECrim allows the interception of electronic communications provided there is prior judicial authorization. The law establishes that the judicial decision must be reasoned, specifying: the subject of the investigation, the period of interception, the reasons justifying the measure, and the conditions of execution and control.

The Constitutional Court, in STC 49/1999, already established that the interception of communications constitutes an interference with the right to privacy and the secrecy of communications, protected by Article 18.3 of the Constitution. In that ruling, the court determined that:

> *Any restriction on this right must be proportionate, necessary, and sufficiently justified by reasons of public interest related to the prosecution of serious crimes. (STS 49/1999, FJ 6.)*

Furthermore, Article 731 bis of the LECrim provides that judges may authorize statements by videoconference when appropriate to avoid unnecessary travel or to protect witnesses and victims of serious crimes. This measure is especially relevant in cases of gender-based violence, human trafficking, or sexual assault, where re-victimization can be avoided through remote statements.

Organic Law 8/2021, on the comprehensive protection of children and adolescents against violence, strengthens the possibility of minor victims of crimes appearing via videoconference to prevent re-victimization, as this guarantee is included in Articles 449 bis et seq. of the Criminal Procedure Act (Ley Criminal).

The use of videoconferences for the statements of those under investigation is regulated in Article 520.3 of the Criminal Procedure Act (Ley Criminal), which allows detainees in penitentiary centers to testify before a judge without the need for physical transfer. This measure has been widely used, especially in cases where the transfer of the accused poses security risks or logistical difficulties.

The globalization of crime has boosted the use of technological means in international judicial cooperation. Spain actively participates in various mechanisms that enable the digital exchange of information in cross-border criminal investigations. Regulation (EU) 2018/1727 of the European Parliament and of the Council on Eurojust has promoted digitalization in judicial cooperation between EU Member States. One of the most significant advances has been the creation of the Eurojust Digital Cooperation Portal, which enables: electronic mutual legal assistance requests, secure exchange of procedural information, and cross-border videoconferencing for taking witness or accused statements in international criminal proceedings.

In this context, the Court of Justice of the European Union (CJEU), in its judgment of October 14, 2021 (Case C-338/20), confirmed that evidence obtained through electronic cooperation between Member States is admissible, provided that the procedural guarantees of the States of origin and destination are respected.

The European Investigation Order (EIO), regulated by Directive 2014/41/EU, has facilitated the collection of digital evidence between Member States. Article 10 of the aforementioned regulation precludes the adoption of any measure or procedure, but establishes a catalog of measures that must in all cases be available in the legal systems of the Member States—the privileged ones—among which is the statement of a witness, expert, victim, investigated person, or accused person, according to Arangüena (2021, p. 377):

> *It should be noted that its execution is mandatory, unless there are reasons for denial, applying the principle of mutual recognition broadly."62 In Spain, this mechanism has been adapted by Law 3/2018, establishing an agile procedure to obtain electronic data stored in other countries.*

At the non-EU level, Spain participates in INTERPOL's I-24/7 system, which enables the real-time transmission of international arrest warrants, alerts on suspected terrorists and organized crime suspects, and the exchange of biometric and DNA databases. In its ruling of July 3, 2022, the National Court analyzed the validity of an arrest made in Spain based on an electronically transmitted INTERPOL Red Notice. In its ruling, the court noted that:

> *International arrest warrants transmitted electronically must be verified to avoid violations of fundamental rights, especially in cases of political persecution.*

3.1.2 Oral Trial Phase

The abbreviated procedure, regulated in Title III of Book IV of the LECrim (Articles 757 et seq.), is the most common in Spanish criminal justice, as it applies to crimes punishable by imprisonment of no more than nine years. In this procedure, the oral trial takes place before the Criminal Court or the Provincial Court (depending on the severity of the crime). On the other hand, the summary procedure, regulated in Articles 622 et seq. of the LECrim, applies to more serious crimes, with imprisonment of more than nine years. In this case, the oral trial takes place before the Provincial Court or the Criminal Division of the National Court, depending on jurisdiction. Some of the digital peculiarities common to these procedures in their oral trial phase are:

- Presentation of digital evidence using LexNet and ACCEDA.
- Videoconference questioning of witnesses and experts. Playback of images and audio in court via electronic means. Use of videoconferencing for taking statements from defendants, witnesses, and experts.
- Incorporation of digital evidence obtained during the investigation, especially in technological crimes and organized crime.
- Use of digital databases and electronic forensic tools for evaluating evidence.

Digital evidence has taken on a fundamental role in modern criminal proceedings, especially in crimes such as computer fraud, economic crimes, cybercrime, and corruption. However, its admission must comply with the guarantees set forth in Article 11.1 of the LOPJ (Job-Laws of the Judiciary), which establishes that:

> *Evidence obtained, directly or indirectly, by violating fundamental rights or freedoms will have no effect.*

Furthermore, as we have already highlighted above, the Supreme Court, in its STS 300/2015, stressed that:

> *To be admissible, digital evidence must comply with the principles of authenticity, integrity, legality, and chain of custody. Any intervention in digital devices must be judicially authorized to avoid the violation of fundamental rights. (STS 300/2015, FJ 2.)*

Finally, in this regard, we must highlight STS 652/2021, of July 22, which states that:

> *The references to presence in the supranational regulations invoked, according to a logical contextual interpretation, cover both in-person modalities: physical or virtual. In neither of these two cases is it correct to speak of the accused's absence. He is present.*

3.1.3 The Criminal Enforcement

The criminal enforcement phase is the final stage of the criminal process, where the final judgment is enforced. Digitalization has significantly transformed this stage, streamlining procedures and improving communication between judicial bodies, penitentiary institutions, and other entities involved.

One of the most significant advances has been the implementation of electronic systems for managing and monitoring the enforcement of sentences and security measures, allowing judges and prosecutors real-time access to information on the execution of sentences. Platforms such as LexNet and the Criminal Enforcement Management System (SIGEP) have optimized the exchange of judicial decisions, release orders, and communications between courts and penitentiaries.

In the prison system, the use of electronic records facilitates the processing of prison benefits, release permits, and grade progressions. Organic Law 7/2021 on Data Protection in the Criminal Field regulates access to this information, guaranteeing the confidentiality of inmate data and the security of its processing.

Furthermore, digitalization has improved the management of alternative sentences, such as permanent location or community service, allowing for remote monitoring of the imposed measures. The use of electronic devices in these cases has proven to be an effective tool for ensuring that sentences are served without the need for imprisonment.

However, digitalization in the criminal enforcement phase also poses challenges, particularly in relation to the right to review automated decisions and interoperability between judicial and penitentiary systems. Ultimately, the digital transformation of criminal enforcement has represented a step forward in terms of efficiency and transparency, although it still requires appropriate regulation to guarantee legal certainty and the rights of convicted people.

3.2 *Protección De Datos Personales En El Proceso Penal*

The protection of personal data in the criminal justice system is crucial to ensuring a balance between effective criminal investigations and respect for citizens' fundamental rights. The collection, storage, and processing of sensitive information, such as criminal records, digital evidence, or intercepted communications, poses challenges in terms of access, preservation, and use.

With the progressive digitalization of the criminal justice system, judicial institutions and law enforcement agencies handle large volumes of personal data on a daily basis. Managing this information presents several key challenges:

- Accessibility and security: It is essential to ensure that data is accessible only to authorized individuals, preventing information leaks or unauthorized access.
- Interoperability between systems: Communication between different platforms, such as the Central Registry of Convicted Persons and Rebels, the Judicial Records System, or police databases, requires secure protocols to prevent privacy breaches.
- Data retention and deletion: It is essential that personal data be retained only for the time necessary to fulfill the procedural purpose and that its deletion be carried out in accordance with the principles of the General Data Protection Regulation (GDPR) and Organic Law 7/2021 on Data Protection in the Criminal Field.

The use of new technologies in criminal justice, such as video conferencing, facial recognition, and data analysis systems, also raises the need for clear regulations that guarantee the right to privacy and the principle of proportionality in the collection and processing of information.

One of the most significant rulings on data protection is the judgment of the Court of Justice of the European Union (CJEU) in the Schrems II case (C-311/18), which had a significant impact on data transfers between the European Union and third countries.

In this ruling, the CJEU invalidated the Privacy Shield, the agreement regulating data transfers between the EU and the United States, finding that it did not offer adequate guarantees of protection against interference by US authorities. In its reasoning, the Court established that:

> *The limitations imposed on the protection of personal data by US regulations are not limited to what is strictly necessary and do not guarantee protection equivalent to that offered in the Union.*

The judgment also underlined the importance of effective judicial control, stating that:

> *The right to effective judicial protection requires that any person whose personal data is transferred to a third country be provided with legal remedies that allow them to assert their rights with guarantees equivalent to those established by EU law.*

In the criminal field, this ruling has direct implications for international judicial cooperation and the exchange of information with third countries. The transmission of sensitive data, such as criminal records or electronic evidence, must comply with the protection requirements imposed by the CJEU to avoid violations of the right to privacy and the principle of proportionality in the collection of evidence.

3.3 Relationship Between Technological Gap and Equality of Arms

The principle of equality of arms is a fundamental pillar of criminal and procedural law, derived from the right to a fair trial enshrined in Article 6 of the European Convention on Human Rights (ECHR). This principle guarantees that all parties involved in criminal proceedings have equal opportunities to present and defend their case without undue disadvantage. However, the increasing digitalization of criminal proceedings poses new challenges to procedural fairness, particularly with regard to access to and use of technology in criminal defense.

Currently, inequality in access to digital media can compromise the effective equality of arms between the prosecution and the defense. Judicial systems increasingly rely on technological tools, from electronic file management platforms to videoconference hearings. This transformation, although beneficial in terms of procedural agility, can generate imbalances when the parties do not have the same technological resources or knowledge to use them. One area where this inequality manifests itself is in the access to and use of digital evidence. The prosecution, in many cases supported by the Public Prosecutor's Office and security forces with advanced resources, has the capacity to obtain and analyze large volumes of electronic data, such as telephone records, intercepted communications, or geolocations. In contrast, the defense may face technical and financial difficulties in conducting counter-expert reports or accessing these same evidence on equal terms.

Furthermore, the digital transformation also impacts particularly vulnerable groups in the criminal justice system, such as those in pretrial detention, those under investigation with limited financial resources, or victims of serious crimes, who may lack the means to interact with digital justice platforms. The lack of adequate access to technology can result in a reduction in the exercise of the right to defense and the ability to actively participate in criminal proceedings.

To correct these imbalances, it is essential to implement measures that guarantee technological accessibility and equity in the criminal process. Some of these measures include:

- Training and technological assistance for legal professionals, ensuring that both the prosecution and defense can effectively use digital media.
- Easier access to technology for defendants and defense attorneys, including the use of designated rooms in prisons for remote connection with the courts and the ability to access digital documentation on an equal footing with the prosecution.
- Protocols for intervention in digital evidence, ensuring that the parties have equal access to the technical means necessary to analyze and refute electronic evidence.

The advancement of digitalization in criminal justice should not entail a setback in procedural equality. It is essential that technological changes be accompanied by measures that guarantee equitable access to technology and the full exercise of the right to defense, preventing the technological gap from becoming a barrier that violates the principle of equality of arms.

4 Proposals for Improving the Digitalization of the Criminal Process: Legislative Harmonization, Technological Interoperability, and Training

The digital transformation of the criminal process has represented significant progress in terms of efficiency and accessibility, but it has also posed challenges in terms of procedural guarantees, interoperability, and training for legal practitioners. For digitalization to be effective and respectful of fundamental rights, it is necessary to implement structural improvements that allow for technological development in harmony with the legal framework.

One of the main challenges in the digitalization of the criminal process is the lack of regulatory harmonization that uniformly governs the use of digital tools in all phases of the procedure. Currently, regulations on digitalization are scattered across various laws and regulations, creating legal uncertainty and difficulties in their application. Furthermore, as we have already demonstrated throughout this work, there is a significant conflict between regulations and recent jurisprudence.

At the European level, Regulation (EU) 2016/679 (GDPR) and Directive (EU) 2016/680 have laid the foundations for the protection of personal data in criminal matters, but their transposition into domestic legal systems still differs among Member States. In Spain, Royal Decree-Law 6/2023 has made progress in the digitalization of criminal justice, but further details are still needed on aspects such as access to electronic files, the preservation of digital evidence, and interoperability between judicial systems.

Regarding technological interoperability, there are currently multiple digital platforms used by the courts (LexNET, ACCEDA, Minerva, Avantius, E-Justicia) that are not always compatible with each other, making the exchange of information between courts and other administrations difficult. To ensure effective digitalization, it is essential to:

- Unify technological standards to ensure interoperability of case management systems.
- Implement solutions based on blockchain and distributed ledger technologies to ensure the integrity and traceability of electronic documents.
- Develop common protocols across the European Union for digital judicial cooperation in criminal matters, facilitating the transmission of electronic evidence and the execution of European Investigation Orders.

Regulatory harmonization and interoperability will enable the digitalization of the criminal process to be not only efficient, but also secure and guarantee protection for all parties involved.

The digitalization of the criminal process cannot be achieved without adequate technological training for legal practitioners. Lack of knowledge in the use of technological tools is one of the main obstacles to the effective implementation of digital justice. Judges, prosecutors, lawyers, attorneys, and judicial officials must be trained to operate new technological systems efficiently and securely.

Currently, many of the difficulties in adopting digitalization stem from:

- Lack of knowledge about the use of electronic platforms (LexNET, Minerva, e-Fidelius, etc.), which leads to procedural errors and delays in processing cases. In practice, many justice officials are unfamiliar with how each of these platforms actually works, which can limit their judicial practice and effectiveness.
- Lack of cybersecurity training, increasing the risk of unauthorized access or leaks of sensitive information.
- Distrust in digital tools, especially in the evaluation of electronic evidence and the use of videoconferencing in criminal trials.

To solve these problems, it is necessary to implement:

- Mandatory and ongoing training plans in digital skills for all legal professionals.
- A manual of good technological practices, with clear guidelines on the presentation of digital evidence, access to electronic files, and information security.
- Creation of technological support teams in courts to facilitate the digital transition and advise judges and prosecutors on the use of new tools.

Adequate technology training will ensure that digitalization does not become a barrier, but rather a tool that optimizes the functioning of the criminal justice system.

5 Conclusions

After this research, we could extract the following conclusions:

I. The digitalization of criminal justice is an unavoidable necessity to modernize the judicial system, adapting it to the demands of an increasingly technological society. This is not merely a technical advance, but a structural transformation that must guarantee greater efficiency, accessibility, and transparency in the exercise of judicial power, while always respecting fundamental rights. However, its implementation is fraught with challenges, especially with regard to ensuring that the fundamental rights of defendants and victims are not violated by the use of new technologies.

II. The regulatory framework for e-Justice Administration in the criminal field has evolved significantly, integrating national and international provisions that promote digitalization and procedural efficiency. However, its implementation must guarantee respect for fundamental rights, information security, and interoperability between judicial systems to avoid inequalities in access to justice. In compliance with these regulations, Spain has adopted measures such as the use of videoconferencing in criminal proceedings, the digitalization of court files, and the interoperability of systems between national and international authorities. However, its implementation continues to pose challenges regarding information security and respect for the procedural rights of defendants.

The Spanish regulatory framework in criminal matters has evolved significantly to adapt to the demands of the digitalization of justice. The recent approval of Royal Decree-Law 6/2023 and Organic Law 1/2025 have consolidated digitalization as a fundamental pillar of the judicial system, with particular emphasis on:

- The mandatory nature of electronic court records.
- Interoperability between agencies and remote access to proceedings.
- The digitization of electronic evidence and communications in criminal proceedings.
- The guarantee of rights in the collection and use of electronic evidence.

Despite advances in digitalization, legislative harmonization and technological interoperability between the various platforms used in the judicial system are crucial.

III. Judicial practice is increasingly digital thanks to various platforms. The increase in the use of LexNet since its introduction is due to a series of interrelated factors, such as the mandatory nature of its use under regulations, the expansion of its capabilities, cultural changes, and adaptation to new technologies, in addition to the impact of the COVID-19 pandemic. Although precise statistics are not available, the platform's constant growth is a clear reflection of its growing importance and relevance in the Spanish judicial system.

The Horus Electronic Judicial Office has established itself as an essential tool for prosecutors in Spain, offering efficient and rapid access to judicial information. Its key features include a court file viewer, power of attorney consultation, and appointment schedule management. This platform has had a considerable impact on the daily practice of legal professionals, contributing to greater efficiency in the judicial system. As new updates are implemented, Horus continues to evolve and improve, providing benefits to both legal practitioners and citizens seeking access to justice.

IV. Interoperability between judicial platforms (LexNet, ACCEDA, Horus, etc.) remains insufficient, creating problems in the exchange of information between courts, tribunals, and public administrations. The lack of uniformity slows down procedures and can undermine the right to a trial without undue delay.

V. However, a significant gap persists between regulation and daily judicial practice, as the implementation of digital technologies is uneven across the country, and their application does not always guarantee full compliance with essential procedural principles, such as immediacy, adversarial proceedings, and the right to defense, generating tensions between legal theory and judicial reality. At the jurisprudential level, the Supreme Court and the Constitutional Court have addressed in various rulings the legality of the use of videoconferencing, digital evidence, and the interception of electronic communications, establishing clear limits for the protection of fundamental rights. There is a real discrepancy between regulations in positive law and jurisprudence, which reveals different realities in the Courts and Tribunals.

VI. The use of videoconferencing in criminal proceedings constitutes one of the most significant technological advances in the digitalization of justice in recent years. However, its widespread implementation requires deep critical reflection, as it directly affects essential procedural principles that guarantee the quality and legitimacy of criminal trials.

While it is indisputable that videoconferencing has made it possible to ensure the continuity of judicial activity in emergency situations—such as during the COVID-19 health crisis—its widespread use as an ordinary method of conducting judicial proceedings poses serious risks to the preservation of fundamental rights. Among these, the principle of immediacy stands out, in particular, which requires the judge or court to personally witness the taking of evidence, assessing not only the content of the statements but also the attitude, gestures, and body language of the participants.

Immediacy is not an empty formality, but a substantive guarantee that allows the court to conduct a comprehensive assessment of the credibility of the parties, witnesses, and experts. Replacing physical presence with a telematic connection can weaken this direct perception, affecting the assessment of evidence and, therefore, the proper formation of the court's judgment. Supreme Court doctrine, in rulings such as Supreme Court Judgment 331/2019, has emphasized that the use of videoconferencing is valid, but it must guarantee the authenticity and effectiveness of the communication and preserve the full scope of the right to defense.

Furthermore, the use of videoconferencing poses specific challenges in terms of party identification and communication security. The lack of secure biometric verification mechanisms or the use of general platforms, such as Cisco Webex, has raised legitimate concerns about the authenticity of the person appearing, the confidentiality of attorney-client communications, and the protection of personal data. Without the implementation of robust systems designed specifically for the judicial environment, the risk of breaches that could affect the validity of procedural actions increases.

Another critical aspect is the impact on the right to a substantive defense. In traditional courtroom settings, the attorney can interact immediately and privately with their client during the trial. In a virtual hearing, this interaction is hampered, affecting the spontaneity of the defense strategy and, ultimately, the effectiveness of legal representation. Especially in the criminal field, where people's liberty or lives may be at stake, digitalization cannot be considered as implying a reduction in procedural guarantees.

At the regulatory level, although Royal Decree-Law 6/2023 has promoted the preferential holding of virtual hearings, its implementation must be carefully considered. The law allows exceptions for serious crimes, but in practice, the judicial authority must expressly justify each decision to hold a procedural act by videoconference, analyzing on a case-by-case basis whether circumstances exist that justify its use without affecting essential rights.

In short, while videoconferencing is a valid and useful tool in modern criminal proceedings, its use should be limited to cases where it does not compromise the principles of immediacy, adversarial proceedings, and the right to defense. It must be avoided that it becomes the general rule of action without proper safeguards, especially in proceedings of special significance or complexity. Digitalization must be accompanied by adequate investment in the judicial system's own technological infrastructure, the adoption of strict identification and security protocols, and intensive training for legal practitioners in the use of these tools.

Technology must serve the process, but never replace or trivialize the guarantees that protect fundamental rights in criminal proceedings. Telejustice must be a quality justice system that fully respects the foundations of due process and ensures that accessibility and efficiency are not achieved at the expense of fairness, transparency, or the material truth of the facts under trial.

VII. The protection of personal data in the criminal justice system is a pillar that must be strengthened in the face of digital advancement. The management of sensitive information, access to criminal records, and the transmission of electronic evidence require strict security protocols and respect for privacy, in line with the General Data Protection Regulation (GDPR) and the European doctrine established in cases such as Schrems II. Fundamental rights, such as the right to due process, the presumption of innocence, the right to defense, and the principle of immediacy, are particularly affected by digitalization. Videoconferencing, for example, can compromise the principle of immediacy if not properly implemented, as reflected in rulings such as STS 331/2019 and STS 652/2021. The use of technologies in criminal proceedings poses significant challenges regarding the protection of personal data and privacy. The management of sensitive information, especially in the digital sphere, requires a robust legal framework that respects the principles of proportionality, necessity, and legitimacy. Case law, such as the CJEU's Schrems II case, underlines the importance of ensuring that the processing of personal data does not infringe fundamental rights.

VIII. The analysis of digitalization in the different phases of criminal proceedings reveals that, while technological advances have made it possible to modernize and streamline the procedure as a whole, the challenges regarding procedural guarantees vary significantly depending on the procedural stage at which digital tools are incorporated.

During the pretrial phase, digitalization has significantly improved access to procedural information through electronic court files, facilitated the management of proceedings, and allowed for more agile transmission of evidence and criminal records, especially in cases involving international cooperation. However, the incorporation of digital investigation techniques, such as the interception of communications or remote access to electronic devices, requires strict judicial oversight to avoid violations of the right to privacy and the confidentiality of communications.

In the oral trial phase, technological innovations such as videoconferencing, the reproduction of digital evidence, and online access to the file have optimized procedural times and facilitated the remote participation of witnesses, experts, and defendants. However, these tools pose serious caveats regarding the principle of immediacy, adversarial proceedings, and the right to defense, especially when a secure identification system or fluid communication between attorney and defendant is not guaranteed.

In the sentence enforcement phase, digitalization has enabled more efficient management of criminal enforcement, facilitating the electronic transmission of decisions, monitoring the fulfillment of sentences, and the online processing of reintegration measures or prison benefits. Despite this, new challenges arise regarding the interoperability of judicial and penitentiary systems, as well as the protection of sensitive personal data.

In conclusion, while digitalization has represented a global advance in the efficiency and accessibility of the criminal process, its practical application must be nuanced and adapted to the specifics of each procedural phase. It is essential that technology be incorporated while scrupulously respecting the principles that govern each stage of the process, always ensuring that innovation does not diminish the essential guarantees that constitute a fair trial. The modernization of the criminal process must be an evolutionary and safeguarding process, combining efficiency with the full preservation of the fundamental rights of the parties involved. The digitalization of the criminal process has clearly improved aspects such as processing agility and administrative transparency, but it has also highlighted new vulnerabilities that must be addressed, such as cybersecurity, the protection of digital files, and the reliability of the presentation of electronic evidence.

IX. It is essential to adopt measures that guarantee equitable access to digital justice for all citizens, especially for those groups in vulnerable situations. The principle of universal access to justice must extend to the digital environment, preventing the lack of technological means from becoming a new cause of inequality. Digitalization has also revealed the existing technological gap between different actors in the criminal process, especially between those who have access to digital resources and those who do not. This can create a disadvantage for the most vulnerable parties, such as defendants in prison or victims, who may not have the necessary tools to interact effectively with the digital system.

X. Artificial Intelligence represents a promising tool, but it also poses serious ethical and legal challenges. Its use should be limited to administrative and logistical support functions, avoiding its automatic application in judicial decisions that affect fundamental rights. Transparency, human supervision, and judicial oversight should be the guiding principles for its integration into criminal proceedings. It must be applied with caution to avoid the risk of discrimination or violation of fundamental rights.

Greater integration of Artificial Intelligence in the analysis and optimization of judicial processes should be promoted, and measures should be implemented to bridge the technological gap and ensure equality of arms between the parties. Furthermore, it is essential to strengthen procedural safeguards in digital justice systems, with particular attention to data protection and transparency in the use of new technologies.

XI. Technological training for judges, prosecutors, lawyers, solicitors, and public officials is essential for true digital justice. The lack of digital skills among legal practitioners not only generates inefficiency, but can also affect the validity of procedural acts and compromise the rights of the parties.

XII. The digital transformation of criminal justice should not be an end in itself, but rather a means to improve the quality of effective judicial protection. The digitalization of criminal justice should not be seen as an isolated or merely technical objective, but as a tool serving a more accessible, efficient, and guaranteed justice system. The process of technological modernization must be intrinsically linked to the protection and reinforcement of the essential principles of criminal proceedings, such as immediacy, publicity, oral proceedings, adversarial proceedings, and the right to defense. These values cannot be sacrificed for the sake of efficiency or procedural speed.

The incorporation of telematic means, such as videoconferencing, electronic court file management, or the use of digital evidence, must always be accompanied by a rigorous analysis of their impact on the fundamental rights of the parties. In particular, it is crucial to ensure that technological advances do not result in impersonal judicial processes, where the judge loses direct contact with the parties, or where the defense finds its capacity for real and effective intervention limited. The guarantee of a justice system that is friendly, humane, and respectful of human dignity must prevail over any automation.

Furthermore, digital transformation must consider reducing inequalities in access to justice as a fundamental objective. A digital justice system that, due to a lack of technological resources or training, excludes certain social sectors, widening the inequality gap, cannot be accepted. Accessibility must not only be physical or economic, but also digital, which implies investments in infrastructure, technological training plans for legal practitioners and citizens, and specific support measures for vulnerable groups. Likewise, digitalization must be accompanied by robust protection of personal data, guaranteeing the confidentiality of procedural communications and the security of information stored and transmitted electronically. Confidence in the administration of digital justice depends largely on the perception that the personal data and privacy of the parties will be protected against breaches and cyberattacks.

Therefore, digital transformation must be guided by a human rights approach, aimed at strengthening the democratic quality of the criminal justice system, consolidating a faster and more efficient administration of justice, but also one that is more fair, accessible, secure, humane, and respectful. Technology must always be an instrument that enhances, not replaces, the basic principles of the criminal process and the essential mission of the judicial system: to guarantee effective judicial protection for all citizens, without exception.

References

Arangüena Fanego, C. (2021), Orden europea de investigación: régimen de sustitución de la medida solicitada. Indret 1, p. 377-401.

CJEU. C-311/18. Judgment of the Court of Justice (Grand Chamber) of 16 July 2020 [request for a preliminary ruling from the High Court (Ireland) — Ireland] — Data Protection Commissioner v Facebook Ireland Limited, Maximillian Schrems

Tribunal Constitucional de España. C. Sentencia del Tribunal Constitucional 49/1999, de 13 de marzo (BOE núm. 65, de 17 de marzo de 1999).

European Convention on Mutual Assistance in Criminal Matters, done in Strasbourg on 20 April 1959

Convention on Cybercrime, done in Budapest on 23 November 2001 (Budapest Convention).

Council Framework Decision 2008/978/JHA of 18 December 2008 on the European Evidence Order for use in criminal proceedings

Real Decreto 1065/2015, de 27 de noviembre, sobre comunicaciones electrónicas en la Administración de Justicia en el ámbito territorial del Ministerio de Justicia y por el que se regula el sistema LexNET. «BOE» núm. 287, de 01/12/2015, Ref. BOE-A-2015- 12999 Entrada en vigor: 01/01/2016

Real Decreto-ley 6/2023, de 19 de diciembre, por el que se aprueban medidas urgentes para la ejecución del Plan de Recuperación, Transformación y Resiliencia en materia de servicio público de justicia, función pública, régimen local y mecenazgo. Jefatura del Estado. «BOE» núm. 303, de 20 de diciembre de 2023. Referencia: BOE-A-2023-25758

Real Decreto-ley 16/2020, de 28 de abril, de medidas procesales y organizativas para hacer frente al COVID-19 en el ámbito de la Administración de Justicia

Díaz, E. (28/12/2023) La digitalización de la justicia obliga a celebrar juicios 'online' a partir de abril. Eleconomista.es. Retrieved from: https://www.eleconomista.es/legal/noticias/12601980/12/23/la-digitalizacion-de-la- justicia-obliga-a-celebrar-juicios-online-a-partir-de-abril.html

Directive 2013/48/EU of the European Parliament and of the Council of 22 October 2013 on the right of access to a lawyer in criminal proceedings and in proceedings relating to the warrant of

Directive 2016/343/EU of the European Parliament and of the Council of 9 March 2016 on the strengthening of certain aspects of the presumption of innocence and the right to be present at the trial in criminal proceedings

Electronic Administration Portal. https://www.administraciondejusticia.gob.es

Constitución Española. Boletín Oficial del Estado, 29 de diciembre de 1978 (CE 1978)

European Convention for the Protection of Human Rights and Fundamental Freedoms (ECHR), done in Rome on 4 November 1950

Consejo General de la Abogacía. https://www.abogacia.es

González Campo, F.A. (2013). Arquitectura de la administración judicial electrónica española: estado actual y perspectivas de futuro. Lasala Calleja, Pilar, ed., Derecho y tecnologías avanzadas, LEFIS Prensas de la Universidad de Zaragoza, p. 36

International Covenant on Civil and Political Rights (ICCPR), adopted by the United Nations General Assembly on 16 December 1966

Poder Judicial. https://www.poderjudicial.es/

Ley 1/2000, de 7 de enero, de Enjuiciamiento Civil. «BOE» núm. 7, de 08/01/2000. Referencia: BOE-A-2000-323

Medina Zepeda, E. (2022). Hacia una teoría sobre la E-justice o justicia digital: instrucciones para armar. Cuestiones Constitucionales. Revista Mexicana De Derecho Constitucional, 1(46), p. 177–212. Retrieved from: https://doi.org/10.22201/iij.24484881e.2022.46.17052

La Moncloa. www.lamoncloa.gob.es

Audiencia Nacional de España. Sentencia de la Audiencia Nacional, Sala de lo Penal, n.º 23/2019, de 20 de noviembre (ECLI:ES:AN:2019:4102)

Ley Orgánica 6/1985, de 1 de julio, del Poder Judicial. Jefatura del Estado. «BOE» núm. 157, de 02 de julio de 1985. Referencia: BOE-A-1985-12666

Ley Orgánica 1/2004, de 28 de diciembre, de Medidas de Protección Integral contra la Violencia de Género. «BOE» núm. 313, de 29/12/2004. BOE-A-2004-21760

Ley Orgánica 1/2025, de 2 de enero, de medidas en materia de eficiencia del Servicio Público de Justicia. «BOE» núm. 3, de 03/01/2025. Referencia: BOE-A-2025-76

Real Decreto de 14 de septiembre de 1882, por el que se aprueba la Ley de Enjuiciamiento Criminal (LECrim). «Gaceta de Madrid» núm. 260, de 17/09/1882. Referencia: BOE-A-1882-6036

European Union Portal. www.eur-lex.europa.eu

Ministerio de la Presidencia, Justicia y las Relaciones con las Cortes. www.mjusticia.gob.es

Rayón Ballesteros, M.C. (2022), Tecnología al servicio del proceso: especial referencia a la celebración de juicios telemáticos, Ius et Scientia, 1, p. 189- 199

Regulation (EU) 2014/910 of the European Parliament and of the Council of 23 July 2014 on electronic identification and trust services for electronic transactions in the internal market and repealing Directive 1999/93/EC (eIDAS Regulation)

Regulation (EU) 2018/1727 of the European Parliament and of the Council of 14 November 2018 on the European Union Agency for Criminal Justice Cooperation (Eurojust) and replacing and repealing Council Decision 2002/187/JHA.

Regulation (EU) 2020/1783 of the European Parliament and of the Council of 25 November 2020 on cooperation between the courts and tribunals of the Member States concerning the taking of evidence in civil or commercial matters (Taking of Evidence) and repealing Regulation (EC) No 1206/2001

Tribunal Supremo de España. Sentencia del Tribunal Supremo 300/2015, de 19 de marzo (ECLI: ES:TS:2023:1012)

Tribunal Supremo de España. Sentencia del Tribunal Supremo n.º 331/2019, de 27 de junio (ECLI:ES:TS:2019:2163)

Tribunal Supremo de España. Sentencia del Tribunal Supremo nº 652/2021, de 22 de julio (ECLI:ES:TS:2021:3144)

Immersive Exhibitions and Cultural Heritage: Legal Tensions, Ethical Challenges, and Strategic Responses

Ana Quintana

Abstract The integration of immersive technologies into museum exhibitions has profoundly reshaped the ways in which cultural heritage is accessed, interpreted, and experienced. Through the use of virtual reality, augmented reality, artificial intelligence, and blockchain, cultural institutions have developed new forms of engagement with historical artifacts and narratives. However, these innovations raise significant legal and ethical challenges, particularly at the intersection of intellectual property law and cultural heritage protection frameworks. This chapter explores the complex relationship between digital innovation and cultural stewardship, analyzing issues such as the legal status of digital surrogates, the commercialization of public domain works, the rights of source communities, and the regulatory implications of emerging technologies. It highlights the tensions between exclusive proprietary claims and collective heritage obligations, underscoring the risk of digital enclosure and cultural commodification. Drawing on comparative legal analysis, international instruments, and best practice frameworks, the chapter offers strategic recommendations for museums, content creators, and policymakers. These include the adoption of open access policies, the integration of ethical and participatory curatorial models, the proactive management of technological and legal risks, and the promotion of transparency, inclusivity, and public accountability. Ultimately, the chapter argues that the sustainable future of immersive exhibitions lies not only in technological advancement but in the cultivation of ethical, legal, and social frameworks that prioritize cultural rights, historical integrity, and public engagement.

Keywords Immersive exhibitions · Intellectual property law · Cultural heritage protection · Digital surrogates · Artificial intelligence · Blockchain · Museum ethics

---

A. Quintana (✉)
Madrid, Spain
e-mail: aquintan@nebrija.es

L. A. Garcia-Segura (ed.), *Collaborative Insights on Technology and Legal Frameworks*, Advanced Sciences and Technologies for Security Applications,
https://doi.org/10.1007/978-3-032-27714-5_5

1 Introduction

The emergence of immersive exhibitions in museums has fundamentally transformed the ways in which cultural heritage is accessed, interpreted, and experienced. Technological advancements such as virtual reality, augmented reality, three-dimensional modeling, and projection mapping have empowered museums to offer visitors highly interactive and multisensory engagements with historical artifacts and narratives. These developments mark a significant departure from traditional museological practices centered on static observation, inviting audiences to become active participants in dynamic cultural environments. However, the increasing reliance on digital reproductions of cultural objects within immersive exhibitions has introduced intricate legal questions, revealing a complex and often underexplored tension between intellectual property law and cultural heritage protection law.

Historically, cultural heritage has been conceived as a collective asset, belonging to humanity as a whole and transcending individual proprietary claims. Nevertheless, the digital replication of cultural artifacts blurs the boundaries between public patrimony and private ownership. The legal status of these digital surrogates, particularly their eligibility for intellectual property protection, remains contentious. Intellectual property law, with its emphasis on incentivizing creativity and rewarding originality through exclusive rights, frequently collides with the objectives of cultural heritage law, which prioritize preservation, authenticity, and universal access. These conflicting logics become especially pronounced when museums, technology companies, or individual creators seek to assert proprietary claims over digital representations of artifacts traditionally situated within the public domain.

Central to this emerging legal landscape is the issue of originality. Copyright law generally conditions protection on the existence of a "personal intellectual creation" by the author. The requirement of originality presents particular challenges when applied to exact or highly faithful digital reproductions of cultural heritage objects. Jurisprudence varies considerably across jurisdictions. In the European Union, the Court of Justice has affirmed in cases such as *Infopaq International A/S v. Danske Dagblades Forening* (Court of Justice of the European Union 2009) that originality necessitates the expression of free and creative choices, potentially excluding mere reproductions from protection. By contrast, certain national legal systems exhibit greater flexibility, sometimes granting "thin" copyright to works that involve substantial technical skill even in the absence of overt creativity. These divergences create legal uncertainty, especially in the transnational context in which many immersive exhibitions are conceived and disseminated.

Beyond questions of copyrightability, the role of cultural heritage law adds further complexity. International instruments, including the * (UNESCO 1970) Convention on the Means of Prohibiting and Preventing the Illicit Import, Export and Transfer of Ownership of Cultural Property* (UNESCO 1970) and the * (UNESCO 2003) Convention for the Safeguarding of the Intangible Cultural Heritage* (UNESCO 2003), establish robust obligations on states to protect and disseminate cultural heritage. National legislations often complement these obligations by restricting the reproduction, modification, or commercialization of certain cultural artifacts. These regulatory frameworks seek to safeguard the integrity and accessibility of cultural heritage, potentially limiting the extent to which museums and private actors may exert proprietary control over digital surrogates.

The commercial dynamics of immersive exhibitions further exacerbate these tensions. While museums have historically operated within a public service ethos centered on education and conservation, the rise of the "experience economy" has incentivized the development of immersive projects with significant revenue-generating potential. Partnerships with private technology firms and the deployment of proprietary platforms have resulted in exhibitions that are not merely educational but also commercial in nature. Licensing arrangements, paywall restrictions, and exclusivity agreements governing access to digital cultural heritage raise profound ethical concerns regarding the commodification and enclosure of the digital commons.

Compounding these issues is the inherent complexity of immersive exhibition projects, which typically involve a multitude of contributors including curators, historians, digital artists, software developers, and institutional partners. The collaborative nature of these projects demands sophisticated contractual frameworks to allocate rights appropriately, address questions of joint authorship, moral rights, and derivative works, and manage the use and reuse of content across different platforms and jurisdictions. Moreover, the increasing incorporation of user-generated content into immersive exhibitions adds another layer of legal complexity, raising questions about consent, ownership, and the scope of permissible uses.

Emerging technologies such as artificial intelligence further challenge traditional legal paradigms. AI-assisted techniques in the reproduction and reconstruction of cultural artifacts complicate notions of authorship and originality, potentially displacing human creators from the center of the legal framework. In parallel, blockchain-based solutions and the proliferation of non-fungible tokens (NFTs) introduce new models of authentication, ownership, and transfer of digital cultural assets, operating largely outside existing regulatory structures and raising new questions about provenance, authenticity, and access.

From a normative perspective, the risk that digital reproductions of cultural heritage might become enclosed within proprietary frameworks is a cause for concern. If access to digital surrogates of public domain artifacts becomes restricted through intellectual property claims, licensing barriers, or technological protection measures, the fundamental values underpinning cultural heritage preservation—namely accessibility, education, and collective memory—may be undermined. The transformation of cultural heritage into a commodity available only to those who can afford access would represent a profound shift away from the ideals of universality and inclusivity that have historically animated cultural policy.

In light of these developments, this chapter seeks to examine the evolving legal relationship between intellectual property law and cultural heritage protection as it manifests in the domain of immersive exhibitions. It will first explore the legal status of digital reproductions of cultural heritage objects, analyzing how standards of originality and authorship are applied across different jurisdictions and legal traditions. Building on this foundation, the chapter will analyze the legal ecosystem of immersive exhibitions, focusing on the allocation and management of rights among multiple stakeholders and the contractual models typically employed to structure such projects. Subsequently, the discussion will turn to the regulatory tensions between intellectual property rights and cultural heritage obligations, illustrating these conflicts through case studies and highlighting the risks of digital enclosure.

The chapter will then present best practices for museums, content creators, and policymakers aiming to navigate these challenges while upholding both the incentives for technological innovation and the principles of cultural stewardship. Particular attention will be given to licensing strategies that promote openness, the responsible use of emerging technologies, and the need for ethical frameworks to guide the digitization and dissemination of cultural heritage. Finally, the chapter will reflect on emerging trends, including the implications of artificial intelligence, blockchain authentication, and evolving audience expectations, concluding with recommendations for future research and policy development.

By undertaking this analysis, the chapter aspires not merely to clarify the existing legal framework but to contribute to a broader reflection on how law should adapt to ensure that technological innovation in the cultural sector serves the public interest. It advocates for an approach that reconciles the protection of intellectual creativity with the fundamental imperatives of preserving, disseminating, and democratizing access to the cultural heritage of humanity.

2 The Legal Nature of Digital Cultural Heritage Replicas

The process of creating digital replicas of cultural heritage objects has become increasingly central to contemporary museum practices, particularly in the context of immersive exhibitions. Through the use of advanced technologies such as three-dimensional scanning, photogrammetry, and sophisticated computer modeling, cultural institutions are now capable of generating highly accurate and detailed digital versions of historical artifacts, archaeological sites, and architectural monuments. These digital surrogates serve multiple purposes: they facilitate preservation efforts by reducing the need for handling fragile originals; they democratize access to heritage by reaching global audiences; and they provide the raw material for the construction of immersive experiences that seek to revitalize public engagement with history, art, and culture.

However, the legal characterization of these digital replicas remains contested and deeply ambiguous. Central to this ambiguity is the question of whether a digital reproduction of a cultural heritage object qualifies as an original work of authorship under intellectual property law, specifically copyright law. Traditionally, copyright protection is premised upon the criterion of originality, a standard that requires the work to be the result of the author's own intellectual creation, reflecting a certain degree of personal input, creativity, and expressive choice.

The European approach to originality, as articulated by the Court of Justice of the European Union (CJEU) in a series of landmark cases—notably *Infopaq International A/S v. Danske Dagblades Forening* (Court of Justice of the European Union 2009), *Painer v. Standard Verlags GmbH* (Court of Justice of the European Union 2011), and *Football Dataco Ltd. v. Yahoo! UK Ltd.* (Court of Justice of the European Union 2012)—underscores that originality exists where the author has made free and creative choices that reflect their personal touch. The emphasis is on creative freedom rather than technical skill or laborious effort. A purely mechanical reproduction of an existing work, no matter how technically sophisticated or demanding, lacks the requisite element of creative freedom and thus does not attract copyright protection. Consequently, in the European Union, a faithful digital reproduction of a public domain artifact—such as a Renaissance sculpture or a medieval manuscript—is generally considered to fall outside the scope of copyright protection.

This principle serves a crucial public policy function: it preserves the integrity of the public domain. Cultural artifacts that have long since passed into the public domain are understood to belong collectively to humanity. They constitute a shared cultural patrimony that should not be re-appropriated through the creation of digital replicas that, if protected by new copyrights, could restrict access and reintroduce proprietary barriers. The risk of what some scholars have called "digital enclosure" is particularly acute in the context of immersive exhibitions, where access to cultural experiences can be controlled, monetized, and subjected to restrictive licensing terms.

Nevertheless, the assessment of originality is not always straightforward. The process of creating a digital replica often involves a range of decisions that could, in some circumstances, be characterized as creative. Choices regarding the method of scanning, the treatment of lighting and texture, the resolution of incomplete or damaged artifacts, and the digital restoration of missing elements all involve degrees of human judgment. In some cases, especially where the digital reconstruction significantly departs from a mere mechanical copy and reflects interpretative or restorative decisions, the threshold of originality may arguably be met.

Comparative law reveals important divergences in this regard. In the United States, while the principle of minimal creativity established in *Feist Publications, Inc. v. Rural Telephone Service Co.* (1991) requires more than mere effort or labor, courts have historically recognized that certain types of photographs, even those aiming at faithful reproduction, may be eligible for copyright protection if they involve elements of creativity in aspects such as framing, lighting, or composition. By analogy, a digital replica that involves creative decisions about the mode of representation could potentially qualify as a protectable work. However, the scope of protection in such cases would be narrow and would not extend to the artifact itself, only to the particular expression embodied in the digital file.

An additional layer of complexity arises from the fact that digital replicas are frequently embedded within larger immersive projects that combine multiple layers of creative and technical contributions. In such contexts, the individual digital replica might be but one element among many in a composite work that qualifies as an original creation overall. The assembly of replicas, the design of the virtual environment, the integration of soundscapes, interactive elements, and narrative structures all contribute to the originality of the immersive experience as a whole, even if individual replicas themselves do not independently satisfy the originality criterion.

Beyond copyright considerations, other legal doctrines may also influence the protection and management of digital cultural heritage replicas. The European Database Directive (Directive 96/9/EC) provides sui generis protection for databases where there has been substantial investment in the obtaining, verification, or presentation of data. A digital archive of cultural heritage replicas could potentially benefit from such protection, although it would not confer rights over the individual replicas per se. Similarly, contractual arrangements, rather than statutory intellectual property rights, often play a critical role in governing access to and use of digital replicas, with museums imposing terms of use that restrict reproduction, modification, or commercial exploitation, even in the absence of formal copyright protection.

Alongside intellectual property frameworks, cultural heritage protection laws impose independent obligations on the treatment of digital surrogates. International conventions, notably the * (UNESCO 1970) Convention* and the *2003 Convention for the Safeguarding of the Intangible Cultural Heritage* (UNESCO 1970, 2003), affirm the duty of states to protect cultural heritage and to promote its dissemination for the benefit of humanity. National laws frequently regulate the reproduction and commercialization of heritage objects, especially when they have been officially designated as protected cultural property. Such regulations may impose restrictions on the ways in which digital replicas can be created, displayed, or commercialized, particularly where issues of authenticity, integrity, or cultural sensitivity are at stake.

The ethical dimensions of digital replication also warrant attention. While digital technologies offer unprecedented opportunities for conservation and education, they also raise questions about authenticity, authorship, and cultural appropriation. The act of digitally reproducing a cultural artifact is not neutral; it involves interpretative decisions that can shape public understanding of the artifact's meaning and significance. In particular, the reproduction of artifacts from indigenous cultures or postcolonial contexts must be approached with heightened sensitivity to issues of representation, consent, and cultural sovereignty.

Given these legal and ethical complexities, museums and cultural institutions must approach the creation and use of digital cultural heritage replicas with a rigorous and informed methodology. Legal due diligence should be conducted to ascertain the intellectual property status of the original artifact, the originality of the digital replica, and any applicable cultural heritage protections. Institutions should implement clear and transparent policies governing the use of digital surrogates, ensuring that access is aligned with their educational mission and public service responsibilities.

Where possible, institutions are encouraged to adopt open licensing models that facilitate the broad dissemination and reuse of digital cultural resources. The use of standardized licenses, such as Creative Commons, can strike a balance between recognizing the contributions of digital creators and preserving the public's right to access and engage with cultural heritage. Moreover, collaborations between museums, technology providers, and source communities should be structured to reflect principles of equity, transparency, and respect for cultural diversity.

As immersive exhibitions continue to expand in scope and ambition, the legal frameworks governing digital replicas of cultural heritage will need to evolve to meet new challenges. Ensuring that the digital transformation of the cultural sector advances the goals of preservation, accessibility, and inclusivity, rather than merely serving commercial interests, will require ongoing vigilance, interdisciplinary dialogue, and thoughtful legal innovation.

3 Immersive Exhibitions: A New Legal Ecosystem

The incorporation of immersive technologies into museum exhibitions has given rise to an entirely new legal ecosystem, characterized by the convergence of multiple fields of law, the participation of diverse stakeholders, and the emergence of complex contractual structures. Unlike traditional exhibitions, which primarily involved the curation and display of physical artifacts, immersive exhibitions operate at the intersection of intellectual property law, contract law, cultural heritage regulation, technology law, and even data protection law. The complexity of this legal environment demands a sophisticated and interdisciplinary approach to rights management, governance, and compliance.

At the heart of this new ecosystem lies the hybrid nature of immersive exhibitions themselves. An immersive exhibition is not simply a collection of objects but rather an integrated experience that combines visual, auditory, tactile, and sometimes even olfactory stimuli to create a multisensory environment. This environment is often built upon a foundation of digital assets, including high-resolution images, three-dimensional models, animations, soundscapes, video sequences, and interactive software elements. Each of these components may be subject to distinct intellectual property rights, and their aggregation into a unified exhibition experience generates additional layers of legal complexity.

The multiplicity of rights involved in an immersive exhibition requires careful identification and management. Copyrights, neighboring rights, database rights, design rights, and trademark rights may all be implicated, depending on the nature of the content and its mode of use. In many cases, the same digital asset may be subject to overlapping rights held by different entities. For example, a 3D scan of a sculpture may be protected by copyright (if sufficiently original), but the software used to create and display the scan may be protected separately under copyright or patent law, and the interface design may attract additional protection under design rights. Furthermore, the museum's branding associated with the exhibition may invoke trademark considerations.

The collaborative nature of immersive exhibition projects further compounds these challenges. Typically, such projects involve multiple stakeholders, including museums, technology providers, creative studios, software developers, sound designers, historians, curators, and often, third-party rights holders. The contributions of these actors must be coordinated through contractual arrangements that define the allocation of rights, the scope of permitted uses, the duration of licenses, the attribution of authorship, the management of moral rights, and the mechanisms for dispute resolution. These contracts form the backbone of the legal infrastructure supporting the exhibition.

One of the critical contractual issues that arise in this context is the allocation of intellectual property rights between the commissioning institution (often the museum) and the creators of the digital content. Museums may seek to acquire full ownership of commissioned works to ensure maximum control over their use and dissemination. However, creators, particularly independent artists and studios, may resist full assignment and instead prefer to grant limited licenses, preserving their own rights to reuse or adapt the content for other projects. Negotiating these arrangements requires balancing the interests of all parties and ensuring that the resulting agreements provide sufficient clarity to avoid future disputes.

Another important aspect of the contractual framework relates to warranties and indemnities. Given the complex rights environment, museums must often rely on representations and warranties from content creators that the works supplied are original, do not infringe third-party rights, and are free from encumbrances. Indemnification clauses are typically included to allocate the risk of infringement claims, although their enforceability and practical utility may vary depending on the jurisdictions involved and the financial standing of the contracting parties.

The integration of user-generated content into immersive exhibitions introduces additional legal considerations. Increasingly, immersive exhibitions invite visitors to interact with digital environments, contribute their own content, or customize aspects of the experience. Such participatory elements raise questions about the ownership of user contributions, the scope of permissible uses by the institution, and the application of data protection laws where personal data is collected, stored, or processed. Clear terms of use and privacy policies are essential to govern these interactions and to ensure compliance with applicable legal standards, such as the General Data Protection Regulation (GDPR) in the European Union (European Parliament and Council, 2016).

Data protection and privacy issues are particularly salient in the context of immersive technologies. Devices such as VR headsets, motion sensors, and biometric trackers may collect sensitive personal data, including facial features, voice patterns, body movements, and even emotional responses. The processing of such data must be carefully managed to ensure that it is lawful, transparent, and proportionate. Museums must implement appropriate technical and organizational measures to safeguard data security and must provide visitors with clear information about the nature of the data collected, the purposes of processing, and the rights available to data subjects.

In addition to intellectual property and data protection issues, the regulation of cultural heritage imposes further constraints on the design and operation of immersive exhibitions. Where digital reproductions of protected cultural artifacts are used, museums must ensure compliance with national and international heritage laws. This may involve obtaining permissions from competent authorities, respecting conditions relating to the integrity and authenticity of reproductions, and avoiding the commercialization of sacred or sensitive cultural objects without appropriate consent (UNESCO 1970, 2003).

Furthermore, ethical considerations play a crucial role in shaping the legal ecosystem of immersive exhibitions. Museums are increasingly expected to adhere to principles of cultural sensitivity, inclusivity, and social responsibility in their programming. The depiction of cultural heritage must avoid reinforcing stereotypes, misrepresenting historical narratives, or appropriating cultural expressions without respect for originating communities. Legal compliance alone is insufficient; institutions must strive to align their practices with broader ethical standards recognized by the international museum community, such as those articulated in the ICOM Code of Ethics for Museums (ICOM (International Council of Museums) 2017).

The funding structures of immersive exhibitions also contribute to the complexity of the legal environment. Public–private partnerships, grant funding from cultural agencies, sponsorships by private corporations, and crowdfunding campaigns each introduce their own legal considerations, including contractual obligations, intellectual property ownership, brand management, and compliance with public procurement regulations. Navigating these frameworks requires a comprehensive understanding of the interdependencies between legal rights, financial arrangements, and institutional missions.

Given the intricate and multidimensional nature of immersive exhibitions, museums and their partners must adopt proactive and holistic legal strategies. Early engagement with legal counsel, careful mapping of rights and obligations, the adoption of clear and fair contractual templates, and the implementation of rights management systems are all critical measures for ensuring legal security and operational success. Moreover, institutions must cultivate a culture of legal and ethical awareness among their staff, recognizing that the deployment of immersive technologies is not merely a technical endeavor but a profoundly cultural and normative one.

As the use of immersive technologies continues to expand, the legal ecosystem surrounding museum exhibitions will likely become even more complex. New challenges will arise from the increasing convergence of technologies, the globalization of cultural exchanges, and the evolving expectations of audiences. Museums must therefore remain vigilant, adaptable, and committed to balancing innovation with the enduring principles of cultural stewardship, public accessibility, and respect for human dignity.

4 Regulatory Tensions Between Intellectual Property and Cultural Heritage Law

The growing adoption of immersive technologies in the cultural heritage sector has not only expanded the reach and impact of museums and exhibitions but has also heightened the tensions between the regimes of intellectual property law and cultural heritage protection. These tensions are neither incidental nor easily resolved;

they reflect deep-seated differences in the underlying objectives, legal philosophies, and societal values that animate the two fields. As digital technologies continue to reshape the landscape of cultural production, dissemination, and consumption, the legal frictions between exclusive rights and collective heritage obligations are becoming increasingly salient.

One of the most visible manifestations of these tensions concerns the digital reproduction of artifacts that reside in the public domain. Cultural heritage objects—paintings, sculptures, manuscripts, monuments—that have passed into the public domain represent a collective patrimony that should, in principle, be freely accessible to all. The digital reproduction of such works presents both an opportunity and a risk. While digitization enables unprecedented global access and facilitates new forms of engagement with heritage, it also opens the door to the re-privatization of public domain works through the assertion of new intellectual property claims over digital surrogates.

When museums or private actors claim exclusive rights over digital reproductions—whether based on alleged originality, technological investment, or through the imposition of contractual terms—they effectively create new enclosures around previously open resources. This phenomenon has been criticized by scholars and policymakers alike as a form of "cultural lock-up," which undermines the public's right to access, study, and reuse cultural heritage materials. As pointed out by Frischmann (Frischmann 2005), the value of public domain resources lies precisely in their openness and the opportunities they provide for downstream innovation, education, and cultural enrichment.

The legal permissibility of asserting rights over digital reproductions varies by jurisdiction, as previously discussed, but the ethical implications are clear: institutions that benefit from public support and hold cultural objects in trust for society must be cautious not to undermine the public's rights in the name of financial sustainability. The principle of "open access" to cultural heritage, endorsed by organizations such as UNESCO (UNESCO 2003) and the European Commission, calls for minimizing barriers to the use and reuse of cultural content, particularly for educational, scholarly, and non-commercial purposes.

Beyond the status of digital reproductions themselves, further tensions arise when immersive exhibitions reinterpret, modify, or fictionalize cultural heritage materials. While creative reimaginations can provide valuable new perspectives and foster emotional connections with audiences, they may also risk distorting historical narratives, misrepresenting cultural values, or trivializing sacred traditions. The (UNESCO 2003) Convention for the Safeguarding of the Intangible Cultural Heritage explicitly warns against the "disfigurement, commodification, and misappropriation" of living cultural expressions (UNESCO 2003). Immersive exhibitions that prioritize spectacle over authenticity may inadvertently violate these principles, even if they remain within the bounds of formal intellectual property law.

This challenge is particularly acute in the case of indigenous and marginalized communities. Cultural artifacts originating from colonial contexts, sacred objects, or traditional cultural expressions are not mere historical relics; they are living embodiments of identity, memory, and spiritual significance. The reproduction and commercial exploitation of such materials without the free, prior, and informed consent of source communities contravenes not only ethical norms but also emerging legal standards recognized in instruments such as the United Nations Declaration on the Rights of Indigenous Peoples (UNDRIP) (United Nations 2007). Articles 11 and 31 of UNDRIP affirm the rights of indigenous peoples to maintain, protect, and control their cultural heritage, knowledge, and expressions.

The ethical tensions surrounding the reproduction and commercialization of indigenous heritage have been increasingly acknowledged in the cultural sector. Initiatives such as the Traditional Knowledge (TK) Labels developed by the Local Contexts project offer practical tools for communicating the appropriate use of indigenous materials in digital environments (Contexts n.d.). Museums engaging in immersive projects must consider incorporating such frameworks to ensure that the rights and wishes of source communities are respected.

Further complications arise from the use of emerging technologies such as blockchain and non-fungible tokens (NFTs). While blockchain offers promising solutions for documenting provenance and ensuring the authenticity of digital cultural assets, its application in the cultural heritage sector raises new regulatory questions. For example, the "tokenization" of digital replicas of cultural artifacts through NFTs can create artificial scarcity around works that are, or should be, freely accessible. The sale of NFTs linked to public domain artifacts by private individuals or commercial platforms risks undermining the public's perception of cultural heritage as a collective good and could lead to disputes over authenticity, ownership, and ethical use.

Additionally, artificial intelligence (AI) systems are increasingly being used to create derivative works based on cultural heritage materials, including automatic reconstructions, stylistic reinterpretations, and generative artworks. The legal status of AI-generated outputs remains unsettled in most jurisdictions, particularly in relation to authorship, originality, and copyrightability. The use of AI to reinterpret or remix cultural heritage raises profound questions about the preservation of authenticity, the protection of cultural values, and the attribution of creative responsibility. It also challenges existing intellectual property frameworks that were not designed to accommodate non-human creators.

The increasing commercialization of immersive exhibitions further exacerbates the tensions between intellectual property rights and cultural heritage obligations. Revenue models based on pay-per-view access, subscription services, merchandise sales, and exclusive licensing agreements create incentives to restrict access to cultural content and prioritize profitability over public service. While museums must navigate economic realities and seek financial sustainability, they must also guard against the erosion of their public trust mandate. Cultural institutions occupy a unique position of authority and responsibility; their actions shape public perceptions of cultural ownership, value, and accessibility.

Recognizing these challenges, a number of cultural institutions and international organizations have begun to articulate principles and best practices for managing the intersection of intellectual property and cultural heritage in the digital age.

UNESCO's Recommendation concerning the Promotion and Use of Multilingualism and Universal Access to Cyberspace (2003) calls for the promotion of equitable access to cultural content in digital formats. Similarly, the International Council of Museums (ICOM) has emphasized the need for museums to balance intellectual property interests with the principles of cultural stewardship, inclusivity, and respect for human dignity (ICOM (International Council of Museums) 2017).

In practical terms, several strategies can be employed to mitigate regulatory tensions. Museums and cultural institutions can adopt open access policies that provide free and unrestricted access to digital reproductions of public domain artifacts, particularly for non-commercial and educational uses. Where intellectual property rights are asserted over digital surrogates, licensing terms should be designed to maximize accessibility, transparency, and fairness. Contracts with content creators should include provisions that recognize and respect the rights of source communities, and collaborative curatorial practices should be encouraged to ensure that multiple perspectives are represented.

Moreover, ethical review processes should be integrated into the design and implementation of immersive exhibitions. Such processes can help identify potential risks related to cultural sensitivity, historical representation, and community engagement early in the project lifecycle. Institutions should also engage in ongoing dialogue with stakeholders, including source communities, audiences, and peer organizations, to refine their practices and adapt to evolving societal expectations.

Ultimately, navigating the regulatory tensions between intellectual property law and cultural heritage protection requires more than technical legal compliance; it demands a broader commitment to cultural ethics, social responsibility, and the public interest. Immersive technologies offer extraordinary opportunities to expand the reach and impact of cultural heritage, but they also carry significant risks of commodification, exclusion, and misrepresentation. Institutions that embrace a holistic and principled approach to rights management, community engagement, and cultural stewardship will be better positioned to harness the transformative potential of immersive exhibitions while honoring their obligations to the past, the present, and future generations.

5 Best Practices for Museums and Content Creators

In light of the regulatory tensions previously outlined, the necessity for clear and responsible best practices in the creation, management, and dissemination of digital cultural heritage content has become increasingly urgent. Museums, cultural institutions, and content creators must navigate a complex and evolving landscape where intellectual property rights, cultural heritage obligations, ethical considerations, and technological innovations intersect. Establishing robust best practices is essential not only to ensure legal compliance but also to uphold the broader responsibilities of cultural stewardship, public trust, and social inclusion.

The first and perhaps most fundamental principle for institutions and creators engaged in immersive exhibitions is the commitment to respect and preserve the public domain. Public domain artifacts represent a collective inheritance that transcends national boundaries, private ownership, and commercial interests. Accordingly, where cultural heritage objects have fallen into the public domain, digital reproductions of those objects should, as a general rule, be made freely available for educational, scholarly, and non-commercial uses. Museums should avoid asserting exclusive rights over faithful digital surrogates, particularly when no creative additions justify a claim to originality. Adopting open access policies, inspired by initiatives such as the Europeana Public Domain Charter (Europeana 2011) or the Smithsonian Open Access project (Smithsonian Institution, n.d.), can help ensure that cultural heritage remains accessible and usable for the benefit of all.

In cases where digital reproductions involve significant creative contributions—such as digital reconstructions, restorations, or artistic interpretations—institutions may legitimately assert intellectual property rights. However, even in these cases, licensing terms should be designed to balance the interests of creators with the public interest. Creative Commons licenses, particularly those that allow for non-commercial reuse with attribution (e.g., CC BY-NC), offer flexible mechanisms to encourage broad dissemination while preserving recognition of the creators' contributions. Where possible, institutions should prefer licenses that minimize restrictions on reuse, such as CC BY or CC0, particularly for content derived from public domain sources (Creative Commons n.d.).

Transparent attribution practices represent another key best practice. Visitors, researchers, and users of digital cultural content should be provided with clear information regarding the provenance, authorship, and licensing status of digital works. Proper attribution not only respects the moral rights of creators but also reinforces the authenticity and reliability of cultural resources. Institutions should ensure that metadata accompanying digital reproductions includes detailed credits, including acknowledgment of source communities where relevant.

Engaging with source communities in the digitization and presentation of cultural heritage materials is essential, particularly where artifacts originate from indigenous peoples, colonized populations, or marginalized groups. Best practices in this area demand early and meaningful consultation with representatives of these communities, the incorporation of their perspectives into curatorial narratives, and the establishment of protocols for the respectful handling of culturally sensitive materials. Collaborative models, such as co-curation, joint intellectual property management, and participatory exhibition design, can serve to redistribute authority and foster a more inclusive and equitable approach to cultural heritage stewardship (Contexts n.d.).

Museums and content creators should also be attentive to the ethical dimensions of digital restoration and reconstruction. When digitally restoring damaged or incomplete artifacts, it is critical to maintain transparency about the interventions made, distinguishing clearly between original elements and reconstructed or interpretative additions. Techniques such as visual markers, layered reconstructions, or annotated metadata can help audiences understand the extent and nature of digital modifications. In doing so, institutions uphold their responsibility to preserve historical authenticity and prevent the unintentional distortion of cultural narratives.

Another important area for best practices concerns the management of user-generated content within immersive exhibitions. As visitors increasingly contribute to digital experiences—through annotations, personal narratives, creative reinterpretations, or interactive responses—institutions must implement clear policies governing the ownership, licensing, and use of such contributions. Terms of use should be drafted in accessible language, specifying whether users retain ownership of their contributions, whether they grant licenses to the institution, and how their data will be handled in compliance with applicable data protection laws such as the General Data Protection Regulation (GDPR) (European Parliament and Council 2016).

Data privacy and security must be prioritized in all projects involving the collection, processing, or storage of personal information. Museums deploying technologies such as VR headsets, motion capture systems, biometric sensors, or AI-driven personalization must ensure that personal data is collected only for legitimate purposes, minimized to the extent necessary, stored securely, and processed transparently. Privacy-by-design and privacy-by-default principles should guide the technological architecture of immersive exhibitions. Institutions should conduct Data Protection Impact Assessments (DPIAs) where appropriate and provide clear, user-friendly privacy notices to all participants.

The ethical deployment of emerging technologies such as AI and blockchain also requires careful consideration. When using AI tools to create or augment cultural heritage content, institutions should critically assess issues of bias, authenticity, and accountability. For instance, AI-driven reconstructions of incomplete artifacts must be presented with caveats regarding their speculative nature, avoiding the illusion of historical certainty where none exists. Blockchain applications in provenance tracking or rights management must be implemented with attention to environmental sustainability, transparency, and accessibility, ensuring that technological solutions do not exacerbate existing inequalities or introduce new barriers to cultural participation.

Financial sustainability strategies for immersive exhibitions should be aligned with ethical and cultural stewardship principles. While commercial partnerships, sponsorships, and revenue-generating activities are legitimate, they should not compromise the integrity, accessibility, or authenticity of cultural heritage presentations. Institutions should ensure that commercial arrangements are structured to protect curatorial independence, avoid conflicts of interest, and prioritize the public interest over purely commercial considerations. Clear labeling of sponsored content, transparent disclosure of financial relationships, and the maintenance of editorial control are critical to preserving institutional credibility and public trust.

Internal capacity building is another crucial element of best practices. Museums and cultural institutions must invest in the training and professional development of their staff to ensure that legal, ethical, and technological competencies are fully integrated into institutional practice. Interdisciplinary teams combining expertise in curatorship, intellectual property law, digital technologies, ethics, and community engagement are essential for the successful design and execution of immersive exhibitions. Continuous reflection, evaluation, and adaptation of practices in response to technological developments and societal changes should be embedded within institutional cultures.

Finally, institutions should actively participate in the broader cultural sector dialogue regarding the responsible use of immersive technologies. Sharing experiences, developing common standards, contributing to policy discussions, and collaborating with peer institutions can help to build a more coherent, ethical, and sustainable ecosystem for digital cultural heritage. Initiatives such as the Europeana Network Association, the UNESCO Memory of the World Programme (UNESCO 2015), and the WIPO Traditional Knowledge Division (WIPO (World Intellectual Property Organization) 2010) provide valuable platforms for collaboration, knowledge exchange, and advocacy.

In conclusion, best practices for museums and content creators engaged in immersive exhibitions must be grounded in a holistic understanding of their multifaceted responsibilities. Legal compliance, while necessary, is not sufficient. Institutions must embrace a proactive commitment to cultural ethics, public accessibility, historical authenticity, technological responsibility, and community empowerment. By doing so, they can harness the transformative potential of immersive technologies to enrich public engagement with cultural heritage while safeguarding the rights, values, and dignity of all stakeholders involved.

6 Emerging Trends and Future Challenges

As museums and cultural institutions continue to expand their engagement with immersive technologies, the legal, ethical, and operational frameworks that govern their activities must evolve to meet the demands of a rapidly transforming landscape. Technological innovation, shifting societal expectations, and emerging regulatory developments are creating new opportunities but also posing unprecedented challenges. Understanding these emerging trends is essential for institutions that aspire to remain at the forefront of cultural stewardship while navigating the complexities of the digital age.

One of the most significant technological developments influencing the future of immersive exhibitions is the rise of generative artificial intelligence (AI). AI systems capable of producing high-quality images, 3D models, soundscapes, and narrative content are increasingly being integrated into cultural heritage projects. AI can assist in the digital reconstruction of damaged artifacts, simulate historical environments, and personalize visitor experiences based on individual preferences. However, the deployment of generative AI introduces complex questions concerning authorship, originality, and authenticity. When AI algorithms generate reconstructions or reinterpretations of cultural heritage objects, determining who—or what—should be recognized as the "creator" of the resulting work becomes problematic.

Most legal systems currently condition copyright protection on human authorship, meaning that purely machine-generated works may fall outside the scope of protection unless human creative input can be demonstrated (WIPO (World Intellectual Property Organization) 2021). This legal uncertainty complicates the management of rights associated with AI-generated content and may expose institutions to unforeseen risks. Furthermore, AI systems trained on existing cultural datasets may inadvertently reproduce biases, inaccuracies, or culturally sensitive material without appropriate contextualization, raising ethical concerns about representation, respect, and historical fidelity.

Parallel to the rise of AI is the growing influence of blockchain technology and the tokenization of digital cultural assets through non-fungible tokens (NFTs). NFTs offer a mechanism for asserting ownership, provenance, and authenticity of digital files, including digital replicas of cultural artifacts. Some museums have experimented with NFT sales as a means of fundraising, expanding audience engagement, and exploring new models of digital ownership. However, the use of NFTs in the cultural heritage sector is fraught with potential pitfalls. Tokenizing public domain works risks creating artificial scarcity around resources that should remain freely accessible. Moreover, the environmental impact of blockchain technologies, particularly those reliant on proof-of-work consensus mechanisms, raises serious sustainability concerns that must be weighed against potential benefits (European Parliament and Council 2022).

The legal status of NFTs remains unsettled, with courts and regulators in different jurisdictions offering divergent interpretations regarding their classification as property, securities, or intellectual property assets. Museums venturing into the NFT space must therefore exercise extreme caution, ensuring that their activities are consistent with both existing legal obligations and broader cultural ethics. Transparent communication with audiences, clear delineation of what ownership of an NFT entails (and does not entail), and alignment with open access principles where appropriate are crucial elements of responsible engagement with this technology.

Another emerging trend concerns the growing expectation of participatory, co-created cultural experiences. Audiences increasingly seek not only to consume cultural content passively but also to contribute to its creation, interpretation, and dissemination. Immersive exhibitions that incorporate user-generated content, crowdsourced narratives, or interactive co-curation models align with these participatory trends but also introduce new legal complexities. Institutions must navigate issues of intellectual property ownership, licensing, consent, privacy, and liability associated with user contributions. Developing clear frameworks that respect user rights while safeguarding institutional interests will be essential for the sustainable integration of participatory practices into immersive exhibitions.

On the regulatory front, governments and international bodies are beginning to pay closer attention to the implications of digital transformation for cultural heritage protection. Initiatives such as the UNESCO Recommendation on the Ethics of Artificial Intelligence (UNESCO 2021) and the European Commission's proposal for an Artificial Intelligence Act (European Commission 2021) reflect growing efforts to create normative frameworks that balance innovation with human rights, cultural diversity, and ethical integrity. Similarly, the expansion of data protection regulations, consumer protection laws, and accessibility standards in digital services will increasingly shape the operational environment for museums engaging in immersive exhibitions.

Cultural institutions must anticipate and adapt to these evolving regulatory landscapes, proactively integrating compliance measures into project planning, risk management, and stakeholder engagement. Failure to do so could expose institutions to legal liability, reputational damage, and loss of public trust. Building internal capacity for legal and ethical risk assessment, participating in sector-wide advocacy for culturally sensitive regulatory frameworks, and fostering dialogue with policymakers will be critical components of institutional resilience.

An additional emerging challenge relates to the globalization of cultural audiences and the resulting transnational legal complexities. Digital and immersive technologies enable cultural heritage content to reach global audiences instantaneously, transcending the territorial boundaries that traditionally structured legal governance. However, this globalization also exposes institutions to a mosaic of conflicting legal regimes concerning intellectual property rights, data protection, cultural heritage restrictions, and consumer rights. Navigating this fragmented legal landscape will require sophisticated strategies, including the adoption of multi-jurisdictional licensing models, the localization of content to comply with national laws, and the negotiation of international agreements to facilitate the lawful exchange of cultural digital resources (Quintana Jiménez 2026).

Cultural sensitivity and inclusivity will also play an increasingly prominent role in shaping the future of immersive exhibitions. As global awareness of issues related to decolonization, indigenous rights, and cultural appropriation grows, institutions will be held to higher standards regarding the representation, interpretation, and stewardship of cultural heritage. Immersive technologies, by virtue of their immersive and often emotional impact, can amplify both the opportunities and the risks associated with cultural storytelling. Institutions must engage in critical self-reflection, consult with diverse communities, and prioritize ethical frameworks that honor the dignity, agency, and diversity of cultural actors (UNESCO 2003; United Nations 2007).

Finally, the evolving expectations of audiences regarding transparency, authenticity, and accountability will demand a rethinking of institutional communication strategies. Visitors will increasingly demand to know not only what is being presented, but also how it was created, whose perspectives are represented, what rights govern its use, and how their own data and contributions are being managed. Institutions that embrace transparency, openness, and dialogue will be better positioned to cultivate trust, foster deeper engagement, and build lasting relationships with their audiences.

In conclusion, the future of immersive exhibitions at the intersection of intellectual property and cultural heritage protection will be shaped by a confluence of technological innovations, regulatory developments, ethical imperatives, and shifting societal expectations. Institutions that approach these challenges with foresight, flexibility, and a steadfast commitment to their public missions will not only navigate the complexities of the digital age but will also help to shape a more inclusive, equitable, and vibrant cultural ecosystem for generations to come.

7 Conclusions and Strategic Recommendations

The integration of immersive technologies into museum exhibitions has introduced profound transformations in the ways cultural heritage is created, interpreted, and experienced. These technological innovations offer extraordinary opportunities for expanding access to heritage, enriching public engagement, and fostering new forms of cultural expression. However, as this chapter has demonstrated, they also give rise to complex legal, ethical, and operational challenges, particularly at the intersection of intellectual property law and cultural heritage protection frameworks.

A fundamental tension persists between the exclusive rights conferred by intellectual property law and the collective obligations underpinning cultural heritage preservation. The digital reproduction of public domain artifacts, the commercialization of cultural content, the deployment of emerging technologies such as artificial intelligence (AI) and blockchain, and the increasingly participatory nature of cultural experiences all amplify the need for institutions to adopt nuanced and ethically grounded approaches to rights management.

At the heart of these challenges lies a critical imperative: to balance innovation and creativity with accessibility, authenticity, and cultural stewardship. Museums and cultural institutions must resist the temptation to prioritize commercial interests at the expense of public trust and historical integrity. Rather, they must embrace their role as custodians of a shared cultural patrimony, ensuring that digital transformations serve to enhance, rather than restrict, the public's engagement with history, identity, and collective memory. This includes safeguarding the values articulated by UNESCO (UNESCO 2003, 2021) and ICOM (ICOM (International Council of Museums) 2017) concerning inclusivity, diversity, and intergenerational access to cultural resources.

Based on the analysis undertaken in this chapter, several strategic recommendations can be proposed to guide museums, content creators, and policymakers in navigating this evolving landscape:

First, institutions should adopt comprehensive open access policies for digital reproductions of public domain artifacts. These policies should minimize unnecessary restrictions on reuse, particularly for educational, scholarly, and non-commercial purposes. By applying standardized licensing frameworks such as Creative Commons (e.g., CC0 or CC BY), museums can promote legal clarity while maximizing cultural reach. Institutions must also ensure that digital surrogates of public domain works are not re-privatized through artificial barriers such as paywalls or restrictive end-user license agreements (Europeana 2011).

Second, museums must integrate rigorous ethical frameworks into the design and implementation of immersive exhibitions. This includes respecting the rights and perspectives of source communities, ensuring transparency in digital reconstructions and restorations, and avoiding the commodification or misrepresentation of culturally sensitive materials. The incorporation of tools like Traditional Knowledge (TK) Labels (Contexts n.d.) can provide culturally appropriate protocols for the sharing and attribution of indigenous knowledge and heritage. Collaborative curation models and participatory design processes should be expanded to democratize the representation of cultural narratives, moving beyond institutional control to embrace co-authorship with historically marginalized groups.

Third, institutions should proactively address the legal complexities associated with emerging technologies. The use of AI requires clear institutional policies concerning authorship, originality, and ethical use, particularly when outputs are derived from sensitive cultural datasets. Where AI is employed for reconstruction or interpretation, explanatory notices should be integrated into visitor experiences to prevent the misperception of speculative outputs as historically verified data. Likewise, blockchain-based systems and non-fungible tokens (NFTs) must be implemented cautiously, with attention to the potential creation of artificial scarcity and their environmental footprint (UNESCO 2021). Any application of NFTs in the cultural heritage domain should include robust transparency mechanisms clarifying what rights are associated with token ownership and how those rights align with the public interest.

Fourth, cultural institutions must strengthen their internal capacities in legal, ethical, and technological domains. Building and maintaining interdisciplinary teams that include legal counsel, curators, digital technologists, and community representatives is crucial. Ongoing training programs, knowledge-sharing initiatives, and institutional reflexivity should be embedded into operational culture. Risk assessment mechanisms such as Data Protection Impact Assessments (DPIAs), ethical review boards, and legal audits can provide structured guidance during project development and implementation.

Fifth, institutions should engage in advocacy and sector-wide dialogue to influence the development of regulatory frameworks that balance innovation with cultural rights. By actively participating in national and international forums—such as those led by UNESCO, WIPO, the European Commission, and ICOM—museums can contribute to the articulation of forward-looking policies that reflect the specificities of cultural institutions. These engagements also present opportunities to collaborate on international frameworks for licensing, data governance, and cross-border access to cultural content.

Sixth, transparency, accountability, and inclusivity must be central pillars of institutional practice. Museums must communicate openly with their audiences about the sources, processes, and rights associated with digital content. Provenance information, methods of digital reproduction, licensing conditions, and decisions involving AI or restoration should be made publicly accessible. Moreover, institutions must acknowledge and address historical injustices, such as the legacy of colonial collection practices, and work actively to decolonize their collections and narratives. Initiatives such as inclusive metadata practices, multilingual interfaces, and accessibility standards are essential to ensuring broad and equitable engagement with digital heritage.

Seventh, institutions must invest in long-term digital sustainability and cultural resilience. This includes the responsible preservation of digital assets through the implementation of open standards, interoperable formats, and future-proof storage solutions. Sustainable funding models should be developed that do not compromise ethical and curatorial integrity, ensuring that digitization efforts and immersive technologies remain aligned with public service values rather than transient market dynamics.

In sum, the future of immersive exhibitions and digital cultural heritage depends not only on technological innovation but also—and more importantly—on the ethical, legal, and social frameworks that guide their development. Museums and cultural institutions that navigate these complexities with foresight, responsibility, and a commitment to public service will not only thrive in the digital age but will also help to shape a more just, inclusive, and culturally rich future for all.

References

Local Contexts. (n.d.). Traditional Knowledge (TK) Labels. Local Contexts Initiative. https://localcontexts.org/

Europeana. (2011). Public Domain Charter. Europeana Foundation. https://pro.europeana.eu/files/Europeana_Professional/Publications/Europeana%20Public%20Domain%20Charter.pdf

Frischmann, B. M. (2005). An Economic Theory of Infrastructure and Commons Management. Minnesota Law Review, 89(4), 917–1030.

ICOM (International Council of Museums). (2017). ICOM Code of Ethics for Museums. International Council of Museums.

Court of Justice of the European Union. (2009). Infopaq International A/S v. Danske Dagblades Forening, Case C-5/08.

Court of Justice of the European Union. (2012). Football Dataco Ltd and Others v. Yahoo! UK Ltd and Others, Case C-604/10.

Court of Justice of the European Union. (2011). Eva-Maria Painer v. Standard Verlags GmbH, Case C-145/10.

Quintana Jiménez, A. (2026). Inteligencia artificial en el turismo y sus desafíos regulatorios. Cuadernos de RES PUBLICA en derecho y criminología, (8), 1–16. https://doi.org/10.46661/respublica.13536

UNESCO. (1970). Convention on the Means of Prohibiting and Preventing the Illicit Import, Export and Transfer of Ownership of Cultural Property. United Nations Educational, Scientific and Cultural Organization.

UNESCO. (2003). Convention for the Safeguarding of the Intangible Cultural Heritage. United Nations Educational, Scientific and Cultural Organization.

UNESCO. (2015). Recommendation concerning the Preservation of, and Access to, Documentary Heritage including in Digital Form. United Nations Educational, Scientific and Cultural Organization.

UNESCO. (2021). Recommendation on the Ethics of Artificial Intelligence. United Nations Educational, Scientific and Cultural Organization.

United Nations. (2007). United Nations Declaration on the Rights of Indigenous Peoples (UNDRIP). United Nations General Assembly.

WIPO (World Intellectual Property Organization). (2010). Intellectual Property and the Protection of Traditional Knowledge. World Intellectual Property Organization.

WIPO (World Intellectual Property Organization). (2021). Museums and Intellectual Property: A Practical Guide. World Intellectual Property Organization.

Hybrid Threats, Disinformation and Foreign Interference in the European Commission's Draft Guidelines on Artificial Intelligence: An Approach

Carlos Galán Cordero

Abstract The Guidelines on Prohibited Artificial Intelligence Practices, as defined by Regulation (EU) 2024/1689 of the European Parliament and of the Council of 13 June 2024 laying down harmonized rules on artificial intelligence and amending certain regulations, provide an overview of AI practices that are considered unacceptable due to their potential risks to European values and the fundamental rights of its citizens and thus of societies and states themselves. This paper provides a summary analysis of the rules contained in these Guidelines as they apply to the fight against hybrid threats, disinformation campaigns and the manipulation of foreign information and interference in the states of the European Union.

Keywords Guidelines artificial intelligence · Prohibited AI practices · Hybrid threats · Disinformation · Foreign interference · Information manipulation

1 Introduction

Recently, the European Commission has published **Guidelines on Prohibited Artificial Intelligence Practices** as defined by Regulation (EU) 2024/1689 of the European Parliament and of the Council of 13 June 2024 laying down harmonized rules on artificial intelligence and amending certain regulations (hereinafter RIA).[1]

[1] Communication To The Commission. Approval of the content of the draft Communication from the Commission - Commission Guidelines on prohibited artificial intelligence practices established by Regulation (EU) 2024/1689 (AI Act).

Annex. to the Communication to the Commission. Approval of the content of the draft Communication from the Commission - Commission Guidelines on prohibited artificial intelligence practices established by Regulation (EU) 2024/1689 (AI Act).

C. Galán Cordero (✉)
Director of the Master's Degree in Intelligence Analysis and Cyberintelligence, Nebrija University, Madrid, Spain
e-mail: cgalan@nebrija.es

L. A. Garcia-Segura (ed.), *Collaborative Insights on Technology and Legal Frameworks*, Advanced Sciences and Technologies for Security Applications,
https://doi.org/10.1007/978-3-032-27714-5_6

These Guidelines, as the Commission itself points out, provide an overview of AI practices that are considered unacceptable because of their potential risks to European values and the fundamental rights of its citizens and thus of societies and states themselves.

As is well known, the RIA is based on two legal bases: Article 114 of the Treaty on the Functioning of the European Union (the legal basis of the internal market), which underlies the prohibitions contained in Article 5 of the RIA, except as prescribed in Article 16 of the same legal body (the legal basis of data protection), which serves as the legal basis for the specific rules on the processing of personal data in relation to the prohibition of the use of remote biometric identification systems for law enforcement purposes, biometric categorization systems for law enforcement purposes and individual risk assessments for law enforcement purposes.

Although the RIA distinguishes between different categories of operators in relation to AI systems: suppliers, implementers, importers, distributors and product manufacturers, the referred Guidelines focus only on suppliers[2] and implementers,[3] given the scope of the practices prohibited in the aforementioned Article 5 of the RIA, directly applicable as of February 2, 2025.

As is well known, the RIA, which aims to promote innovation while ensuring high levels of protection of health, safety and fundamental rights, classifies AI systems into different risk categories: those prohibited, those of high risk and those subject to transparency obligations, and although the aforementioned Guidelines have been drafted to ensure a consistent, effective and uniform application of the RIA across the EU, they are not binding, so that the only authoritative interpretation will be that emanating from the Court of Justice of the European Union (CJEU). In any case, to flesh out the legal area of the Guidelines, the Guidelines provide legal explanations and practical examples to assist interested parties in understanding and complying with the requirements of the RIA.

On the other hand, and since AI is proving to be a particularly effective tool for generating hybrid threat attacks, disinformation campaigns and actions of foreign interference and manipulation of information (baptized by the European External Action Service as FIMI[4]), the purpose of this paper is to present an overview of how

---

[2] According to the provisions of Article 3.3 of the RIA, natural or legal persons, public authorities, agencies or other bodies that develop AI systems or have them developed and place them on the Union market, or put them into service under their own name or trademark, are to be considered ***suppliers***. For their part, suppliers established or located outside the Union are subject to the provisions of the RIA if they market or put into service such systems in the Union, or if the output of the AI system is used in the Union.

[3] According to the above precept, ***implementers*** are considered to be those natural or legal persons, public authorities, agencies or other bodies that use AI systems under their authority, unless the use is for a personal non-professional activity, it being understood that the "authority" over an AI system assumes responsibility over the decision to implement the system and over the manner of its actual use. These implementers are within the scope of the RIA if their place of establishment or location is within the Union or, being located in a third country, if the output of the IA system is used in the Union.

[4] *Foreign Information Manipulation and Interference* (FIMI). The EEAS defines FIMI as a pattern of behavior that threatens or has the potential to adversely affect political values, procedures and

the aforementioned Guidelines can also be used to deal with these threats, some of the most important that beset the citizens, societies and states of the European Union.

It should be recalled at this point that Article 2.3 of the RIA states that this rule does not apply to areas outside the scope of application of Union law and must not, in any case, affect the competences of the Member States in matters of national security,[5] regardless of the type of entity to which the Member States have entrusted the performance of tasks in relation to those competences, expressly excluding from its scope of application AI systems that are "placed on the market, put into service or used with or without modification exclusively for military, defense or national security purposes, irrespective of the type of entity carrying out those activities, put into service or used with or without modification exclusively for military, defense or national security purposes, irrespective of the type of entity carrying out those activities", whereby the application of this exclusion depends on the purposes or uses of the AI system, and not on the entities carrying out the activities with that system, which may also include private operators entrusted by Member States with the performance of tasks in relation to those competences.

As the published text states, the Commission will formally adopt the Guidelines when they are available in all language versions, at which time they will become applicable.

2 Hybrid Threats, Disinformation and FIMI in EU Guidelines

The Guidelines address several important issues related to foreign disinformation and interference in information systems in the context of the use of artificial intelligence systems.

The most significant of these guidelines are summarized below:

---

processes. Such activity is manipulative in nature and is carried out in an intentional and coordinated manner. The actors in such activity may be state or non-state actors, including their proxies within and outside their own territory.

[5] According to the case law of the Court of Justice of the European Union (6 October 2020, La Quadrature du Net and Others, C-511/18, C-512/18 and C-520/18, EU:C:2020: 791, paragraph 135; Judgment of the Court of Justice of 5 June 2023, Commission v. Poland, C-204/21, EU:C:2023:442, paragraph 318, referring to Case C-439/19, paragraph 67 and Case C-306/21, paragraph 40.), the term "national security" refers to "the overriding interest in protecting the essential functions of the State and the fundamental interests of society, and encompasses the prevention and repression of activities which may seriously destabilize the fundamental constitutional, political, economic or social structures of a country and, in particular, directly threaten society, the population or the State itself, such as terrorist activities".

2.1 Manipulation and Deception Techniques[6]

The Guidelines document exposes how AI systems can be used to deploy subliminal, manipulative or deceptive techniques, capable of distorting perception or altering people's behavior, constituting an essential element of what has been generically called *fake news* or, even worse, of disinformation campaigns orchestrated by third party agents, such as states.

With regard to the prohibition of the use of subliminal techniques, it is worth highlighting some examples pointed out by the Guidelines, which we understand to be clearly applicable to hybrid threats or state-orchestrated disinformation campaigns, such as the following:

- Subvisual and subaudible signals: an AI system can display stimuli that are not only subtle or masked, but are presented in a way that makes them completely undetectable by human senses under normal conditions, e.g., flashing images, too fast for the human eye to consciously detect, or playing sounds at volumes imperceptible to the human ear.
- Embedded images: an AI system can hide images within other visual content that are not consciously perceived, but that the brain can process and influence behavior.
- Deflection: an AI system can draw attention to specific stimuli or content to avoid noticing other content, often by exploiting cognitive biases and vulnerabilities in attention.
- Time manipulation: an AI system can alter the perception of time in user interactions, thus influencing their behavior and causing impatience and dependence.

Given the above, it seems clear that the rationale behind these prohibitions is to protect the individual autonomy and well-being of EU citizens from manipulative, deceptive and exploitative AI practices that may subvert and impair the autonomy, decision-making and free choices of individuals.[7]

2.2 Exploitation of Vulnerabilities[8]

The Guidelines prohibit the use of AI systems that exploit people's vulnerabilities based on their age, disability or specific socioeconomic status, all with the aim of distorting their behavior and potentially causing significant harm.

Although, as with other terms, the RIA does not define the concept of "vulnerabilities," the Guidelines encourage us to understand that it encompasses a broad spectrum of categories, including cognitive, emotional, physical and other forms

[6] In relation to Article 5(1)(a) of the RIA. The detailed explanation of the techniques and their impact can be found in sect. 3.2 of the Guidelines.

[7] Recital (29) of the RIA.

[8] In relation to Article 5(1)(b) of the RIA, as developed in sect. 3.3 of the Guidelines.

that may affect the ability of an individual or group of individuals to make informed decisions or influence their behavior, and although Article 5.1.b) of the RIA refers to *any* vulnerability, it narrows the set of persons protected by the prohibition to those defined by their age, disability or socioeconomic status, who in principle have a more limited ability to recognize or resist manipulative or exploitative AI practices and are in need of greater protection.

2.3 Deep Fakes and AI-Generated Content[9]

The Guidelines note the obligation to label "deep fakes" and other content or text generated through AI applications on matters of public interest.

This measure, like the first one mentioned above, aims to reduce the risk of deception and the distorting effects on the formation of people's opinions and beliefs. However, it is noted that, in some cases, even with proper labeling, misleading techniques can have significant effects and distort people's behavior, which could be covered by the prohibition of Article 5(1)(a) if the conditions set out therein are met.[10]

2.4 Chatbots and Misinformation[11]

The Guidelines remind us that a chatbot can present false or misleading information that influences people's behavior, especially if it is not visibly disclosed that users are interacting with an AI system,[12] which significantly raises the risk of chatbots being used to spread misinformation, individually or orchestrated by third party states.

---

[9] Paragraphs (56), (71) and (90) of the Guidelines. It also relates to Article 5(1)(a) of the RIA, if all conditions for prohibition are met.

[10] "*Article 5. Prohibited AI practices.*

1. The following AI practices are prohibited:

a) the introduction on the market, putting into service or use of an AI system that uses subliminal techniques that transcend the consciousness of a person or deliberately manipulative or deceptive techniques with the purpose or effect of substantially altering the behavior of a person or a group of persons, appreciably impairing their ability to make an informed decision and causing them to make a decision they would not otherwise have made, in a way that causes, or is reasonably likely to cause, substantial harm to that person, another person or a group of persons;"

[11] Chatbots are mentioned as an example in sect. 3.2.2.b) of the Guidelines, specifically in paragraph (82).

[12] As these Guidelines remind us, although, in principle, the transparency obligations of Article 50 of the RIA are intended to minimize the manipulative effects of deep fakes and chatbots, there may be cases and contexts in which, despite information warnings, these deceptive techniques continue to have significant effects on individuals and distort their behavior to the point of undermining individuals' individual autonomy and informed decision making, and therefore should not be misused

2.5 AI Learning of Deceptive Techniques[13]

As is well known, AI systems can learn manipulative or deceptive techniques simply because such behavior can contribute to increasing their performance in the tasks for which they were developed. This learning can even occur without conscious intention on the part of the providers, for example, through so-called *reinforced learning.*

Indeed, whether intended by the provider or not, an AI system can learn manipulative techniques because the data on which it has been trained contains examples of manipulative techniques, or because reinforcement learning from human feedback can be altered by manipulative techniques.

2.6 Biometric Categorization Systems[14]

Biometric categorization systems that infer sensitive characteristics such as race, political opinions, union membership, religious or philosophical beliefs, life or sexual orientation of individuals are prohibited; systems that can be used for discrimination and political manipulation.

As the Guidelines remind us, "*Categorization of an individual by a biometric system is typically the process of establishing whether an individual's biometric data belongs to a group with some predefined characteristic. It is not a matter of identifying an individual or verifying his or her identity, but of assigning him or her to a particular category. For example, an advertising display may show different ads depending on the individual looking at it, based on their age or* gender".[15]

Finally, we should not forget that this prohibition does not apply to the "tagging", filtering or categorization of biometric data sets acquired in accordance with Union or national law, which may be used, for example, for law enforcement purposes.[16]

---

for disinformation and manipulative purposes and could still be covered in some cases by the prohibition in Article 5.1.a), if the other conditions imposed by the prohibition, including significant harm, are met.

[13] This aspect is addressed in sect. 3.2.1.c) of the Guidelines, in paragraph (73).

[14] The prohibition of these systems is found in Article 5(1)(g) of the RIA, which is explained in Sect. 8 of the Guidelines.

[15] Article 29 Working Party, *Opinion 3/2012 on developments in biometric technologies,* WP193, 27.4.2012, p. 6.

[16] Recital (30) of the RIA.

2.7 *Interference or Interference in Democratic Processes*[17]

Although the term "interference" is not explicitly mentioned, the Guidelines implicitly address interference in democratic processes by prohibiting the use of AI to influence people's behavior in a way that undermines their autonomy and ability to make informed decisions, which is particularly relevant in the context of electoral processes and public opinion formation.

On the other hand, and although the RIA also does not define what it calls "intentional manipulation techniques", these tools, in the framework of the risk of foreign interference, should be understood as techniques designed or objectively oriented to influence, alter or control the behavior of individuals in a way that undermines their individual autonomy and free choice, being designed to exploit cognitive biases, psychological vulnerabilities or situational factors that make people more susceptible to being influenced; Moreover, we should not forget that, due to their facility to adapt to changing environments, AI systems are also capable of acting on the circumstances or vulnerabilities of a particular individual and, consequently, increase the effectiveness and impact of large-scale manipulation.

As these Guidelines remind us, although the ability to manipulate is an important element in determining the nature of the technique, it is not necessary that the provider or the implementer or the system itself that implements the manipulation techniques also intends to cause harm.

At this point, it is worth remembering something that the Guidelines point out and that seems important to us in this type of threat: distinguishing manipulation from (legal) persuasion is crucial to delineate the scope of the prohibition of Article 5.1.a) of the RIA, which, as is well known, does not apply to legal practices of persuasion.

Indeed, although both manipulation and persuasion influence people's decisions and behaviors, they differ significantly in terms of methods and ethical implications. Thus, *manipulation* involves, in most cases, covert techniques that, by exploiting psychological weaknesses or cognitive biases, undermine personal autonomy, leading individuals to make decisions that they would not otherwise have made if they were fully aware of all the elements necessary for the proper exercise of judgment.

the contrary, *(legal) persuasion* operates within the limits of transparency and respect for individual autonomy, presenting arguments or information in a way that appeals to reason and despite eventually addressing emotions, provides relevant and accurate information to ensure informed, free and autonomous decision making.

[17] As we have said, the term cited in the Guidelines is not explicitly mentioned, although their clear application, among other instances, in electoral processes can be inferred from their postulates.

2.8 Transparency Obligation[18]

The Guidelines emphasize the importance of transparency in the operation of AI systems and the need to provide clear information on their capabilities and limitations.

The purpose of this requirement is to ensure that system users can make properly informed decisions and avoid being victims of misinformation.

3 Conclusions

Although, obviously, the Guidelines on Prohibited Artificial Intelligence Practices, as defined by the European Artificial Intelligence Regulation, are not specifically aimed at countering the enormous risks currently posed by hybrid threats, most notably state-orchestrated disinformation campaigns and foreign interference and manipulation of information, a scientifically oriented reading allows us to conclude that they can also be used to combat such threats, without this being considered in any way as a closed and unique repertoire.

We should add that the European document itself states that the rules contained in this first version will be gradually supplemented with new proposals to combat the risks that the RIA itself has pointed out in the use of applications based on AI systems.

[18] Transparency obligations are mentioned in several sections of the Guidelines, especially in relation to the measures that providers must take to prevent their systems from being declared as "prohibited for use".

Techniques and Processes of Traditional Criminal Intelligence Analysis

Luigy Martino, Claudio A Payá-Santos, Roger Sanz-González, and Victor Rodriguez-Gonzalez

Abstract This work explores the primary methods used in criminal intelligence, particularly emphasizing the intelligence-gathering techniques defined by UNODC and, for data integration in criminal analysis, the processes of link analysis and flow analysis for crime analysis. As the complexity of criminal networks increases, driven by a hyperconnected world and the development of new technologies that open new options for committing crimes, both physical and digital, law enforcement agencies must work in concert to develop and use more effective techniques.

Keywords Criminal intelligence · Intelligence process · Organized crime · Artificial intelligence

1 Introduction

Criminal intelligence analysis is one of the most important activities for law enforcement and national security agencies. As a process, it provides the tools and frameworks necessary to prevent and fight organized crime, financial crimes, and other

L. Martino
Khalifa University Abu Dhabi, Abu Dhabi, UAE

University of Bologna, Bologna, Italy

L. Martino
e-mail: luigi.martino3@unibo.it

C. A. Payá-Santos (✉) · R. Sanz-González
Valencia International University, Valencia, Spain
e-mail: claudio.paya@professor.universidadviu.com

R. Sanz-González
e-mail: rsanzg@professional.universidadviu.com

V. Rodriguez-Gonzalez
Isabel I University, Burgos, Spain
e-mail: victor.rodriguez.gonzalez@ui1.es

L. A. Garcia-Segura (ed.), *Collaborative Insights on Technology and Legal Frameworks*, Advanced Sciences and Technologies for Security Applications,
https://doi.org/10.1007/978-3-032-27714-5_7

illicit activities. Indeed, this is what happened in the case of criminal intelligence, with multiple agencies, including the United Nations Office on Drugs and Crime and Interpol, contributing with their methodologies.

From a chronological perspective, criminal analysis techniques saw significant development during the years when social network analysis began to spread. To specify, social network analysis is a methodological approach that studies social structures through networks and graph theory. It focuses on mapping and measuring the relationships and interactions between individuals, groups, organizations, or other entities known as "nodes". These nodes are connected by "ties" or "edges", representing their relationships or interactions (Otte and Rokusseau 2002a, 2002b). This framework is particularly suited to organized crime (or OC), to highlight the links between different organizations and between groups and individuals within OC.

The first applications of criminal intelligence dates back to the 1920s and 1930s, when the United States police began systematically collecting information on anarchist movements, and the 1950s, when they started studying mafia associations. Later, in the 1970s, criminal analysis techniques were formalized and refined thanks to the contributions of researchers Harris and Harper from Anacapa Science in Santa Barbara, California. In the following decade, the arrival of personal computers, the development of specific software, and the establishment of the first professional associations in the field contributed to the rapid dissemination of this methodology among police agencies and the techniques associated with it. Among the most important organizations is the "IALEIA", In the 1980s, the IACA, and the LEIU were established in the EEUU.

In the expansion of criminal intelligence, essential initiatives were also adopted by international institutions from the police sector, as well as by the European Union and member-state bodies. The main legal steps can be summarized as follows:

- In 1985, the Committee on Crime of the International Association of Chiefs of Police (IACP) issued a Criminal Intelligence manual among member countries to standardize analytical techniques;
- In Europe, the methodological framework for Criminal Intelligence was introduced in the United Kingdom in the late 1980s, in Belgium in the early 1990s, and shortly thereafter adopted by EU Member States in the "Trevi Group", and in 1996 by the General Assembly of Interpol; The general model, in addition to distinguishing between operational and strategic analysis, includes six phases: planning and goal identification, information collection, evaluation, collation of information, analysis, report, and distribution. The intergovernmental cooperation body, the "Trevi Group," was established in Rome at the end of 1975 with the initial aim of improving cooperation between EU member countries in the fight against terrorism. In 1988, the scope of the Trevi Group was further expanded with the creation of a new working group called "Trevi 1992," tasked with developing police cooperation among Member States in preparation for the application of the Schengen Convention. Among the group's objectives was improving police cooperation in the fight against terrorism or other forms of organized crime. The signing of the Maastricht Treaty and later the Amsterdam Treaty marked the end

of the Trevi Group, as the areas of Justice and Home Affairs (Third Pillar) were transferred to the European Union's institutional framework (Fernández and Liz 2023).
- The signing in 1995 of a convention establishing the European Police Office (Europol) ratified by Law No. 93 of March 23, 1998;
- In 1996, the Council of Europe, based on decisions by the Ministers' Committee of the EU Member States with Recommendation R (96) (8), urged Member States to employ both operational and strategic analysis and electronic systems in the fight against organized crime, drug trafficking, and financial crimes, and to establish specialized police units. Recommendation No. R (96) 8 and 2001 No. 11 in Council of Europe. For the historical roots of Criminal Intelligence;
- In 2002, the Council of Europe initiated a push to standardize the application practices of crime analysis techniques across different police agencies of its member countries;
- The United Nations Office on Drugs and Crime (UNODC) in 2001 and 2011, under a regional justice assistance program for southeast-European countries, disseminated a manual for training analysts based on the intelligence model used in the United Kingdom.

2 What is Criminal Intelligence, and What is It Used for?

Before examining criminal analysis techniques, defining some terms related to intelligence activities applied to criminal phenomena and their underlying processes is essential. The term "intelligence" comes from the Latin *intelligo* (meaning "to choose, select, discern"), and is related to the Greek word *lego*. British culture already used it in the sixteenth century when it referred to methods employed by information experts. Intelligence is also defined as "processed information" capable of providing additional knowledge.

More specifically, the term *intelligence* is defined in literature as a set of activities that include research, data collection, evaluation of sources and information, comparative analysis of available elements, formulation of hypotheses, preparation of a descriptive report, and dissemination of the final document. When applied to the criminal domain, except for a few authors (e.g., Ratcliffe 2007), who differentiate between "criminal intelligence analysis" (which refers explicitly to studying criminal conduct, typically carried out by police personnel) and "crime analysis" (which is used to describe crime patterns and trends and is usually conducted by civilian analysts), criminal intelligence can be defined as:

> *"information that is gathered, analyzed, and shared to anticipate, prevent, or monitor criminal activity. Its primary goal is to collect, evaluate, and distribute information about individuals who are reasonably suspected of being involved in or preparing for criminal activities. The focus of criminal intelligence is on tracking the actions and behaviors of specific individuals or organized groups to gather evidence for prosecution. This type of information is typically not widely distributed and is shared only when necessary" (IACP, 2021), p. 2.*

According to the UNODC, the *intelligence process* consists of five stages: collection, evaluation, collation, analysis, and dissemination, emphasizing analysis, which is, therefore, the focal point of the entire intelligence activity.

Analysis activities are investigative methods based on breaking down a seemingly unified object into its constituent parts, aiming to support the work of specific individuals, such as "investigators, decision-makers, and policy-makers," to "prevent and control crime" (Harris 1978a, b; Di Paolo 2000; Ben-Israel 1989; Payá, 2023).

Regarding different objectives and methods, a distinction is made between "*strategic analysis*" and "*operational analysis*," depending on whether the focus is on general objectives (strategic analysis) or specific cases (operational analysis). Strategic analysis retains a focus on long-term goals, as well as examines current and emerging trends in the criminal environment, such as threats to security and public order (Contreras, 2026; Gangi-Guillén, 2025). While operational analysis studies specific criminal acts, analyzing individuals' or criminal groups' methods, capabilities, and their specific vulnerabilities. The two investigative approaches employ different types of techniques (Caligiuri and Pili 2021; Martino and Merenda 2021).

For now, leaving aside the techniques of operational analysis, which will be discussed in detail later, those related to strategic analysis primarily use statistical tools as they aim to identify trends in criminal phenomena through the study of specific patterns. This approach also foresees the development of descriptive reports on criminal phenomena, which, depending on the purpose, are divided into: a) situation reports, which focus on the current state of crime; b) threat assessments, which consider the potential impact of crime on the territory; c) risk assessments, which focus on the more vulnerable sectors of society where crime might penetrate and exploit these areas to obtain illicit profits (Di Paolo 2000. p. 50–52; Cannavicci 2006. p. 461–464; Stephen and Schneider 1995. P. 403).

While different, the two approaches are interdependent, as each provides the other with essential elements for a better understanding of criminal reality, as emphasized in the United Nations manual:

> *"... it is important to note that a good knowledge of operational intelligence is a highly recommended prerequisite to developing a strategic intelligence capability. The development of operational intelligence in itself will provide an important source of intelligence to consider from a strategic perspective"(p.7).*

The process underlying it can be categorized as cyclical to remain in the scope of operational criminal intelligence activities (see Fig. 1), meaning it is repeated whenever new relevant information is acquired for analysis. Starting from "raw data," the intelligence activity, through the analysis of acquired data, allows the transformation of "raw data into descriptions, and conclusions for intelligence consumers" (Johnston 2003. p. 61–71).

As shown in Fig. 2, the analysis itself encompasses additional activities (represented by the grey background of the image), related to data integration and their interpretation (hypotheses and conclusions). As outlined in its essential elements, the intelligence process is preceded by a series of steps that concern the context,

Fig. 1 The intelligence cycle as per the UNODC criminal intelligence manual for analysts

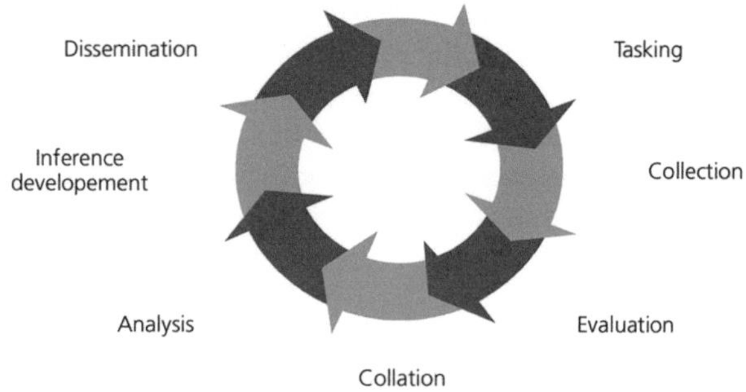

objective, definition of the problem, and the type of information to be collected. More specifically, these can be defined as:

- "context analysis": identifying the environmental variables that characterize a given criminal phenomenon (the Mafia, Drug Cartels, etc.);
- "identification of the objective": the research subject can involve various activities, such as drug trafficking in a specific socio-territorial context, money laundering, or studying criminal organizations (Cosa Nostra in the Italian region of Sicily);
- "problem definition": once the research objective is defined, it is necessary to identify the problem to be analyzed. For example, if the aim is to explore how a particular type of crime spreads within the economic fabric of a given area, one must investigate how organized crime infiltrates the legitimate economy;
- "collection plan": the collection plan must be oriented to the problem being examined and should describe the data to be collected, the offices or entities that need

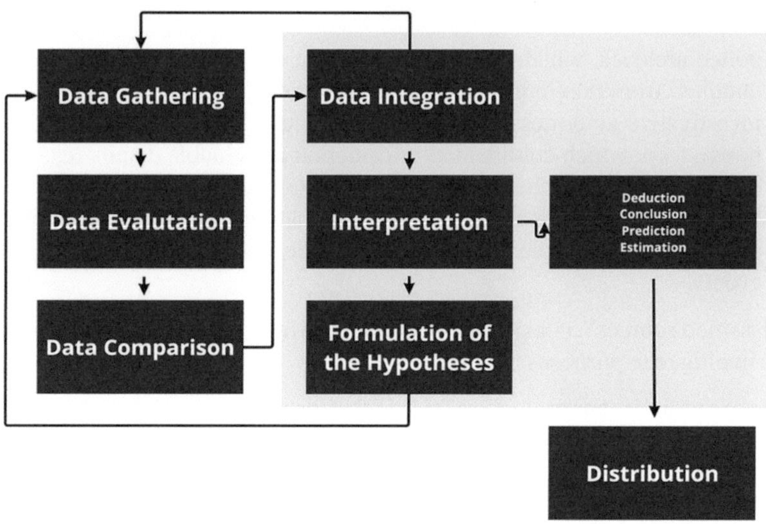

Fig. 2 Intelligence process and analysis

to provide the requested data, a list of the sources from which the data will be gathered, and the timeframe for acquiring the data (Schneider 1995. p. 403–424; Dintino and Martens 1981. p. 58–64; McDowell, Wardlaw and Schmidt 1991. p. 11).
- This intelligence process ends with the distribution or "dissemination" of the processed information to investigative units in order to guide their investigative direction.

3 The Stages of the Intelligence Process

This section will detail the steps of the intelligence process, particularly the techniques used to organize and represent the collected data.

3.1 Information Collection

Once the research subject is defined and the underlying problem identified, it is necessary to gather the information. The UNODC field manual divides the sources that can be used into three main categories based on whether they are accessible to a particular party or not:

a) "open sources", accessible to anyone, such as newspapers, periodicals, telephone directories, referred to in literature as "open source information";
b) "closed sources", accessible only to police forces. Within this classification, we can distinguish the following sources:

   - police archives, which contain information on subjects and criminal events obtained from previous investigative activities;
   - investigative activities (surveillance, searches, wiretaps, undercover operations), from which confidential information or valuable details regarding the examined criminal phenomenon can be obtained;
   - interviews and informal discussions: information gathered by the police informally, such as simple tips citizens provide about individuals and specific events.

c) "classified sources", consisting of confidential information collected exclusively for intelligence purposes.

Most of this information, coming from police investigations and intelligence activities, consists of relational data: individuals in relation to other individuals; individuals with regard to criminal groups; criminal groups with regard to other groups or higher-level criminal networks; individuals, groups, and criminal networks in relation to specific criminal events or complex criminal phenomena (Godfrey, and Harris 1971; McDowell, Wardlaw and Schmidt 1991, p. 22; Izzi, 2011. pp. 23–24).

3.2 Evaluation

Once the data is collected, it is evaluated through a procedure that involves checking the reliability of the sources and assessing the accuracy and validity of the information. This is a crucial activity, as the more reliable the sources and accurate the information, the more consistent the hypotheses formulated at the end of the analysis will be, as the United Nations Office on Drugs and Crime appropriately emphasizes:

> "the Validity of an Inference is Directly Linked to the Quality of the Data Behind the Inference." (2011.P.11).

The way the information is systematically carried out using a procedure known as "*4* × *4*", which assigns a value to the reliability of the source (expressed with a letter of the alphabet) and a value to the content of the information (represented by a number). The source evaluation scheme provided by Europol includes four parameters for the source and four for the information. The source, regardless of its type, is evaluated independently of the information and is commonly classified as follows:

A - a reliable source in terms of authenticity, dependability, and competence (e.g., undercover agent or a source who has consistently proven reliable);

B - a reliable source in most cases;

C - a source that is unreliable in most cases;

X/D - a source whose reliability cannot be assessed (e.g., a new source). The guidelines created by Europol assign to the letter "D" a final code for the evaluation of the source. Previously, the letter "X" was used. The letter "X" is still employed in the field manuals (Di Paolo 2000. p. 37).

The information is classified according to the following numerical parameters:

(a) information whose accuracy is indisputable (because, in addition to coming from a source A, it has found further confirmation);
(b) the information is personally known by the source but not by the agent providing it;
(c) the information is not personally known by the source but is backed by other already recorded information;
(d) the source does not personally know the information and cannot corroborate it in any way. The Drug Enforcement Administration (DEA) uses a more accurate system for evaluating sources and information (6 × 6), based on 6 parameters.

From this evaluation, systemic codes can be created to attest to the source's reliability and accuracy and validity of the information.

3.3 Collation

This is a procedure in which the analyst, selecting relevant data, organizes them into specific categories, for example, by type of criminal group (Cosa Nostra, Camorra,

Colombian cartels), by specialization (drug traffickers, arms dealers), by modus operandi, or by geographical area, etc.

The categories of collated data can relate to any type of entity as long as they serve the purpose of the analysis, such as businesses, specific areas of a city, places where suspicious individuals gather, or biographical details of suspects.

A key activity at this stage of the process is data storage. The programs available today allow for both storage and quick retrieval of data, and especially the possibility of representing the relationships among existing entities (Martens 1987. p. 133).

3.4 *Analysis and Data Integration*

Analysis is the most critical phase of the entire intelligence process. It involves breaking down the unit of analysis into its constituent parts to examine and interpret the acquired data. Essentially, there are two crucial steps in the activity as defined in operational analysis manuals:

(a) the breakdown of the unit of analysis into its constituent elements;
(b) the reassembly of the previously acquired data.

The purpose of the analysis activity is to provide, starting from the collected data, explanations for a given criminal phenomenon by developing certain inferences—essentially, to answer questions regarding the type of illicit activity (drug trafficking versus arms trafficking), the financial flows generated by criminal activities (movement of illegal capital, money laundering), or the territorial scope controlled by the criminal organization.

In the context of operational analysis, if focused on the study of criminal organizations, it aims (also for judicial police purposes) to define the hierarchical structure of the criminal group, interpersonal relationships within the group, and the roles held by the individuals within the criminal association, in order to uncover:

"linkages, associations, and interactions of all known members of an organized crime group with the other organized crime groups" (Ianni, and Reuss-Ianni, 1990. pp. 67–84; Loomis, Palmer and Lupsha 1980. pp. 36–38; Sparrow 1991. p. 255–279; Gonzalez, 2023).

The ultimate goal of this activity is to identify the vulnerabilities of the criminal group in order to dismantle or weaken it. From an operational perspective, the analysis of criminal phenomena, in its organizational forms, involves a dynamic activity that requires the analyst to acquire new information if the initial hypotheses are not supported by reliable evidence, until, by repeatedly going through the analysis cycle, a final inference can be formulated (see Fig. 3):

> "Information Gaps and Weaknesses in the Enquiry, Thus Ensuring that the Analyst Will Continue Data Collection, Even at the Earliest Stages of Analysis Work" (2011. P. 29).

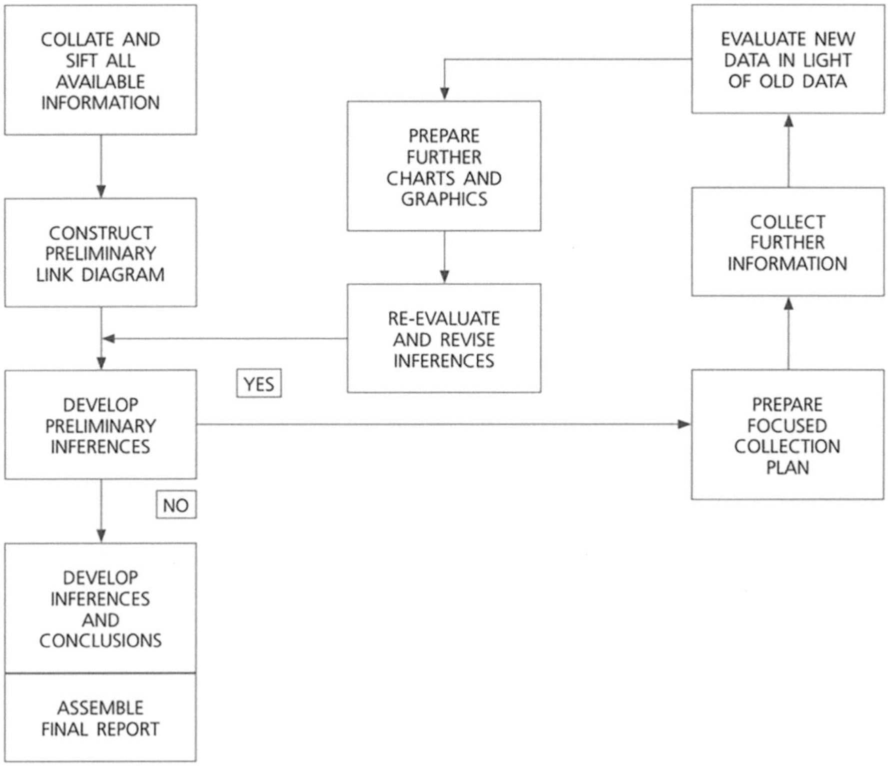

Fig. 3 Detailed process of analysis (United Nations Office on Drugs and Crime, Criminal Intelligence Training, p. 29)

Indeed, if the analyzed phenomena present knowledge gaps, one could use an inductive approach by formulating questions that allow the gathering of more information until a conclusion is reached.

3.5 Data Integration with Criminal Analysis Techniques

Data analysis is a critical moment in the entire intelligence process, as the United Nations manual appropriately highlights:

> *"Data integration is the first phase of the analytical process, combining various types of information from different sources to establish areas of weakness in order to draw inferences for Law Enforcement action".*

Additionally,

> *"Much raw data in an investigation is collated into complex and detailed reports. Other data pertinent to the analysis of the criminal enterprise or suspected criminal activity is*

frequently voluminous, and varied in form. The basic problem for intelligence analysts is putting information together in an organized way so the difficult task of extracting meaning from the assembled information is made easier. Link Analysis puts information about the relationships among entities — individuals, organizations, locations, and so on — into a graphic format and context that will clarify relationships and aid in inference development. Link Analysis can be applied to relationships among those entities, which might have been identified in a given analysis" (2011. p. 26-29).

As the manual emphasizes, data integration primarily involves organizing and representing the data to identify possible relationships between entities (people, organizations, places) and develop possible explanations for the analyzed phenomenon.

For the purpose of this piece, we will consider, due to their similarity to Social Network techniques, the techniques used in *Link Analysis*. These tools are employed to analyze and represent the connections or links between entities or the exchange flows generated by the entities, such as exchanges of goods or information, phone communications, or simply to link criminal events and activities aimed at committing crimes. For analysis techniques, 26 different techniques and 37 products are identified. Most operational analysis techniques are focused on Link Analysis. (Stockley 1988, p. 299).

3.5.1 Link Analysis

Link Analysis is the most widely used approach in operational analysis for data integration. In general, Link Analysis can be defined as a technique focused on studying the interactions between subjects, between subjects and machines, and among the machines themselves. It seeks to understand how different activities or events related to a given process are connected within a particular system.

The first applications date back to the early twentieth century, initially employed in production processes, and later in military communication systems, studying the interactions between pilots and onboard instruments, as well as the analysis of electronic systems. In the realm of production processes, Link Analysis was first applied by Frank Bunker Gilbreth, who, in order to improve industrial production by eliminating downtime, reconstructed the relationships between the activities performed by employees and workers within the company, representing them through a graphical diagram. Frank Bunker Gilbreth (Fairfield, July 7, 1868 – June 14, 1924), was a pioneer of motion economization studies (Thomson 1972. pp. 133–144; Chapanis 1959; Stuster and Chovil 1994. pp. 8–15; Lippert 1971. pp. 269–282) (Fig. 4).

In the field of military intelligence, it was used during World War II by U.S. services for analyzing phone call flows (*communication Link Analysis*), and in the 1960s, by the CIA to reconstruct family ties of groups belonging to clandestine communist organizations in Thailand. Its systematic use in criminal intelligence dates back to the early 1970s, when two researchers from Anacapa Science in Santa Barbara, Harris and Harper, applied Link Analysis to study criminal organizations. In 1975, the two scientists, while developing a training program on Link Analysis

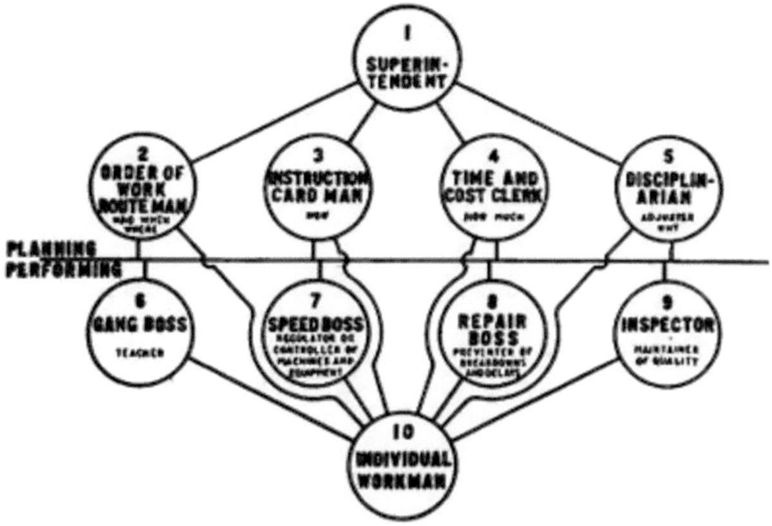

Fig. 4 Graph on the employees' activity of an American manufacturing industry (Gilbreth and Gilbreth 1917, pp. 1–30.)

for police personnel, decided to test this methodology by presenting a real case for examination with this approach (Delgado et al., 2019).

In particular, investigators, divided into twenty-nine groups, were asked to identify the most important figures of the criminal group using this technique and indicate the type of connection (strong or weak) between group members. The goal of the two researchers was to offer Law Enforcement agencies a structured approach to studying organized crime, as they highlight:

> *"Link Analysis procedures were developed and evaluated to aid law-enforcement agencies integrate collected information and develop hypotheses leading to the prevention and control of organized crime. The procedures were designed to portray the relationship among suspected criminals, to determine the structure of criminal organizations, and to identify the nature of suspected criminal activities [...]. A completed Link Analysis might show, for example, the hierarchical structure within a criminal organization, or the interlocking structure of a combination of legitimate and illegitimate organizations"* (Harper and Harris 1975. pp. 157-164).

With this technique, as (Mena, J. 2003.) highlights, it is possible to represent:

> *"Relationship and associations among objects of different types has proven crucial in helping human investigators comprehend complex webs of evidence and draw conclusions that are not apparent from any single piece of information"*(Mena 2003. p. 80).

In light of these definitions, we can assert that this perspective and the techniques underlying it, when applied to criminal phenomena in its organizational forms, allow us to identify:

(a) the presence or absence of connections between subjects and between them and entities of different types (places, organizations, etc.);

(b) how subjects interact within a group;
(c) the central figures within the organization;
(d) the strength of the connections between the subjects belonging to a criminal association (Xu and Chen 2004. pp. 473–487; Schneider 1995. pp. 409–410; Harris 1978a, b. pp. 17–22).

3.5.2 Methodology

According to the Link Analysis approach, from a methodological standpoint, data is organized in an association matrix and represented with a link diagram, both of which are somewhat similar to the sociomatrix and sociogram (Fig. 5a and b).

This allows for identifying four characteristics of a network, which the manuals classify into the following elements:

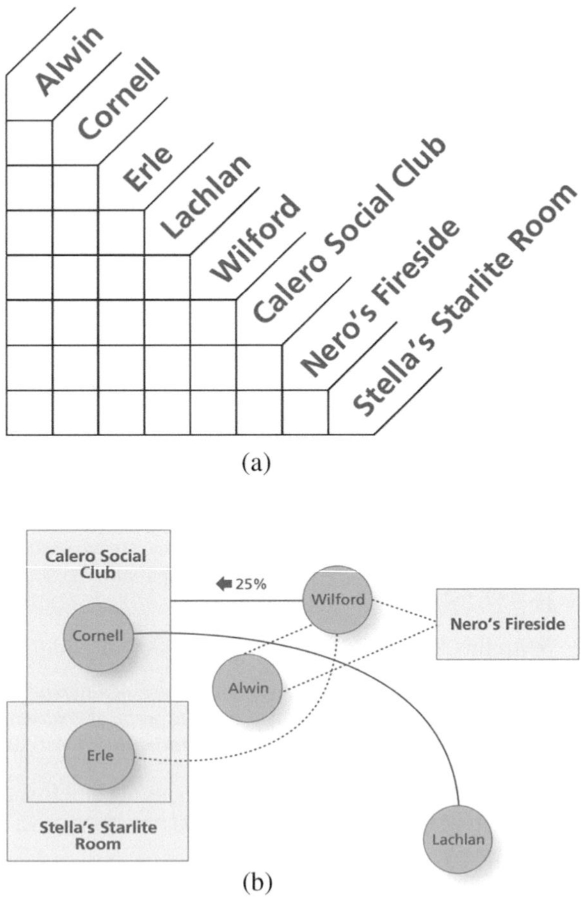

Fig. 5 **a** Instruments of link analysis: sociomatrix. (2011. pp. 45–46) **b** Instruments of Link Analysis: Sociogram. (2011. pp. 45–46)

- *entities*, which are the items under investigation and could include people, businesses, organizations, means of transport, locations, events, objects, etc.;
- *relationships*, which can be familial or based on reciprocity (exchange and compromise), suitability ("the right person to do the job"), bonding (past associations), control (criminal hierarchy or threat), predominance, superiority, subordination, and succession;
- *directionality*, which relates to the flow of information, favors, and authority, and enables understanding of the internal mechanics of a network;
- *strength*, which is a subjective judgment based on interactions included in the relationships and evaluation of the data provided.

What is meant by *entity*? The first type of entity detected using Link Analysis techniques consists essentially of individuals or groups of people (criminal organizations), as operational intelligence analysis primarily aims at identifying the leaders and the type of criminal structure. However, the concept of entity is much broader, encompassing, in addition to people, other elements such as:

- places, such as a physical address or IP address, or certain coordinates like latitude;
- organizations, such as cells, gangs, governments, commercial or military units;
- logistics, such as factories, airports, hotels, schools, warehouses;
- individuals, such as names, titles, or identification numbers;
- components, such as chemicals, fertilizers, acids;
- documents, such as passports, driver's licenses;
- money, such as cash, bank transfers, money orders;
- weapons, such as guns, knives, rifles, bombs;
- vehicles, such as airplanes, trucks, boats, cars.

In relation to entities, understood as individuals, other attributes are also recorded, such as the name, aliases, sex, affiliation, religion, marital status, citizenship, race, date of birth, profession, origin, hair color, eye color, height, weight, and countries visited.

One of the interesting aspects of this perspective, when its primary declared objective is the identification of relationships between entities, is the classification of connections into three types: "strong", "weak", and "absent." The reference to the strength of the connection here is to be understood as the product of the type of interaction between the subjects (friendly, familial) and the evaluation of the information related to that particular interaction. Defined according to the perspective of Link Analysis, the concept of link and entity, it is useful to describe the fundamental steps for data integration. For this purpose, Link Analysis manuals outline a series of steps in which the analyst, after organizing the data into an association matrix, identifies the links between entities, assigning each a value based on the reliability or unreliability of the information underlying the identified relationship, as Harris and Harper clarify:

> *from the intelligence analyst's point of view, the strength of a link between two individuals is a matter of degree, depending upon the content and completeness of available information. At least three categories of links are needed in the application of Link Analysis to criminal*

intelligence: strong links, weak links, and no links. In the definition of strong and weak links, the two scholars add, besides the informational aspect, other characteristics of the link that concern the nature of the relationship and their intensity: "An example of a strong link would be the link between a father and son observed frequently together. A weak link is considered as a relationship that is suspected but unconfirmed, or requires some inferential judgment. An example of a weak link would be a link established on the basis of frequent telephone calls made by an individual to a motel operated by a suspected crime figure" (Harper,and Harris, 1975, p. 58).

In other words, if the link originates from reliable and accurate information, it is considered a strong tie, while in the opposite case, it is considered a *weak tie* (Sparrow 1991, p. 267).

The coding of data in a matrix takes these differences into account, and therefore, from a graphical point of view, relationships are represented with a series of symbols: a solid circle for certain or "strong" links, and an empty circle for presumed or "weak" links.

The same schema is used to represent other types of relationships between entities, such as those between individuals and organizations related to functions, and between subjects concerning financial exchanges. Therefore, the plus sign (+) is used to indicate a subject performing a specific directive function, e.g., within a company, a minus sign (–) indicates a subject performing a presumed directive function, and an arrow is used to indicate whether an investment is confirmed or not (see Table 1).

According to the coding system adopted by criminal intelligence (4 × 4), confirmed links are those evaluated as A1, A2, B1, B2, while unconfirmed links are reviewed in another way (A3, A4, B3, B4, etc.). The subsequent operations involve determining the intensity of the links between entities by summing the number of connections for each entity and their representation (Figs. 6 and 7).

The link graph represents the different coding for confirmed or "strong" links with a solid line, while unconfirmed or "weak" links are shown with a dashed line.

A link diagram aims to represent the relationships between entities. It serves as a starting point for formulating hypotheses regarding the structure of relationships between members of the observed organization, as the United Nations document specifies:

"a chart is a working product from which hypotheses could be gleaned regarding the status of associations among the members of the organization or network. Features of a thinking

Table 1 Symbols of information coding

Association codes	
Code	Meaning
●	Confirmed association between two entities
○	Suspected association between two entities
+	Confirmed member of the organization—officer, manager, employee, club member
—	Suspected membership in the organization
➔	Confirmed investment with no other participation—shareholder, limited partner (direction from owner to owned)
▸	Suspected investment with no other participation (direction from owner to owned)

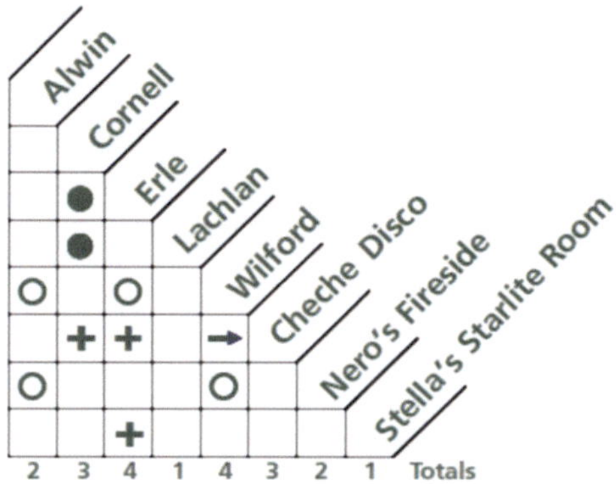

Fig. 6 Insertion of coded data into the matrix

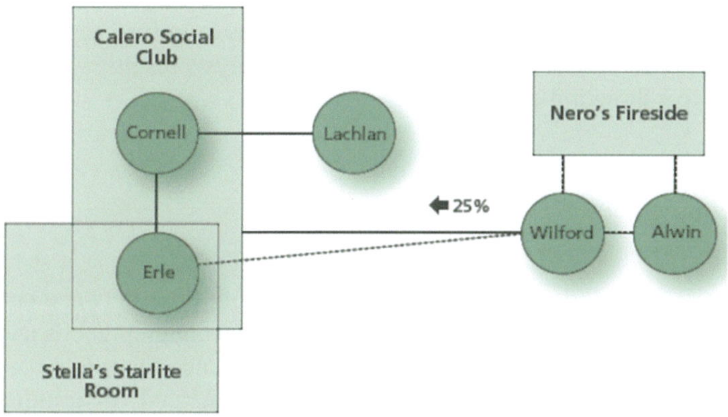

Fig. 7 Link graph

> *association analysis include looking at relationships and links, their strengths and purposes, what those links mean to the organization and to those investigating an organization"* (2011. pp. 45–46).

However, recognizing the static nature of the chart, the document advocates a dynamic approach:

> *"often an association chart shows only a 'freeze-frame' snapshot of the group. It may be more appropriate to show the evolution of the group over time"* (2011. pp. 45–46).

Data representation, therefore, according to this perspective, is the key moment from which to start interpreting the data, formulating a series of questions regarding the structure as a whole, the positions of individuals within the organization, and the structure of the links. Following this classification, the main questions for interpreting a link diagram can be summarized as follows:

- Who are the central figures of the criminal organization?
- What roles do the subjects play within the observed organization?

- What is the nature of the information exchanged among the group members?
- Which communication channels should be monitored?
- How often do group members interact with each other?
- Who are the intermediaries or brokers between different criminal groups?
- How is the hierarchy defined within the group?
- How do the links between entities evolve?

These types of questions relate to the typical inquiries of structural analysis.

3.6 Distribution

The final step consists of disseminating the product(s) of the analysis using reports, summaries, and schematizations from the data that was examined above. Essentially, this final part of the process involves sharing the analysis results, which will guide the investigative direction of the investigative units.

4 Flow Analysis

Criminal analysis techniques, in addition to identifying links between entities, aim to detect communication flows between subjects or exchanging goods and services within a given criminal organization. The same importance for operational analysis is held by the reconstruction of events as they are "generated" by criminal activity over time, with the aim of identifying possible relationships among them. From this perspective, manuals classify flow analysis based on the type of flow examined into:

(a) commodity flow analysis (flow of goods or services);
(b) telephone records analysis (communication flow);
(c) event flow analysis (flow of events).

4.1 Flow of Goods

Most criminal organizations move goods such as drugs, weapons, services, and information to obtain profits. For example, by reconstructing the drug flow, it is possible to determine who is involved in importing, trafficking, retail distribution, and money flow. This allows for identifying the functions and roles held by individual members of the organization (leader, affiliate, etc.) and the type(s) of illicit activity carried out by the criminal group. Knowledge of the flow of goods can also provide the analyst with insights into the structure of the organization, as noted in the manuals:

"In many cases, the flow of the commodity indicates the hierarchical makeup of the organization which in turn leads to an understanding of the power base of the organization" (2011. p. 53).

From an operational standpoint, the flow technique also uses the same tools as Link Analysis: the association matrix, in which the data related to the flow is organized, and the link diagram for graphical representation, consisting of circles for entities and lines for connections. Arrows indicate the direction of the flow (unidirectional and bidirectional) (Di Paolo 2000, pp. 78, 101–103; Schneider 1995, pp. 409–410).

In the matrix shown in Fig. 8a., the connections originating from the exchange of goods between the entities of the group composed of A, B, C, D, and E are indicated, while the link diagram reconstructs the flow of goods originating from A and directed towards B and C, which then, through D and E, return to A.

FROM	TO					TOTAL
	A	B	C	D	E	
A		●	●			2
B				●		1
C						1
D	●				●	1
E	●					1

(a)

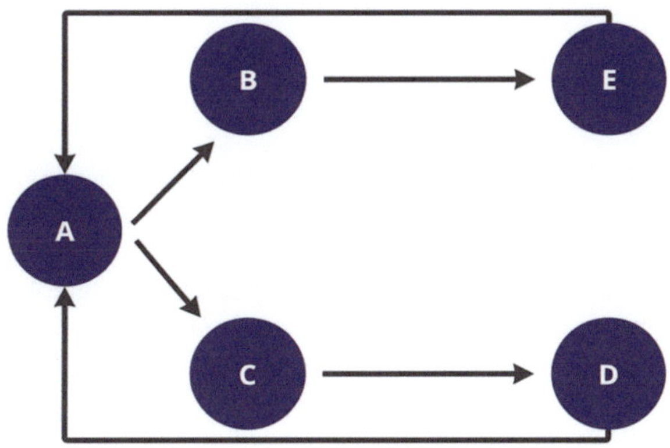

(b)

Fig. 8 **a** Reconstruction of a flow of goods according to the Link Analysis approach: Matrix of flow of goods between entities **b** Reconstruction of a flow of goods according to the Link Analysis approach: flow of goods diagram

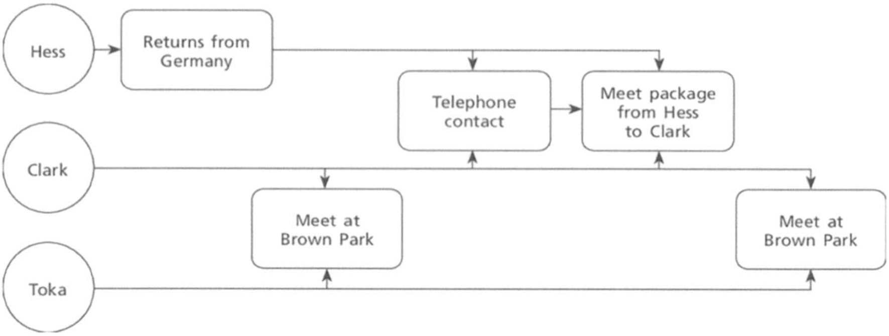

Fig. 9 Example of an event flow diagram

4.2 *Flow of Events*

Event analysis consists of reconstructing the chronological sequence of related events concerning a given criminal phenomenon. To reconstruct an event flow, a link diagram between events is generally used, showing the relationship between them. This type of technique allows the schematization of banking movements, meetings between people, events occurring simultaneously or in rapid succession. In the diagram shown in Fig. 9, the events involving Hess, Clark, and Toka from September 19 to October 5 are represented. In analyzing criminal analysis techniques, we will overlook activity flow analysis since it does not yield significant relational data, as activity flow diagrams are used to organize a set of activities connected by a cause-and-effect relationship aimed at committing a specific criminal event.

4.3 *Telephone Flows*

The analysis of telephone flows reconstructs the telephone contacts between subjects and, mainly, the "frequency" of their communications. It is a technique that can be applied to telephone traffic and any communication (SMS, email, etc.). It also uses the tools of Link Analysis: the association matrix and the link graph, which is used to reconstruct incoming and outgoing telephone traffic and its frequency. This technique is generally employed to determine the nature and frequency of connections between subjects belonging to a criminal group. In telephone flow analysis, the data is entered into a square matrix, where the rows contain the phone numbers from which the calls originate, and the columns contain the recipients' phone numbers. In the graphical representation of the telephone flow, geometric shapes (circles) represent the phone numbers, lines with arrows indicate the connection and the direction of the telephone flow, and the numerical value on the link shows the frequency of the calls (Fig. 10a and b).

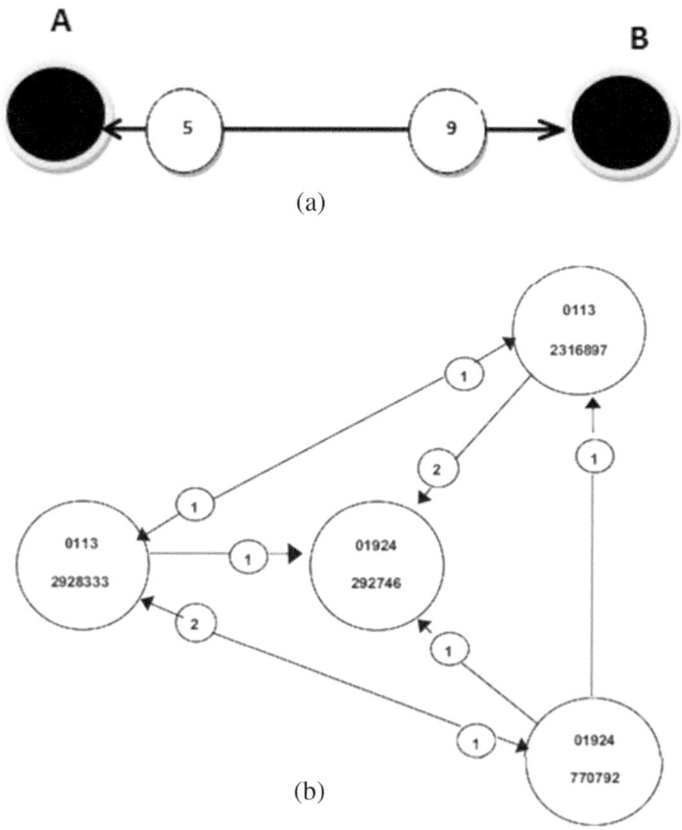

Fig. 10 **a** Examples of telephone flow schematization: telephone traffic between A and B **b** Examples of telephone flow schematization: link and frequency diagram between 4 telephone numbers

An additional example of a telephone flow diagram reconstruction is with a star graph (see Fig. 11), where the telephone flows between a key player and their network, with the leading actor or leader in the center and all the others on the periphery. This type of graph can, of course, be completed with other types of information such as the name of the holder or user of the line (fax, mobile, etc.), their address, and the location of the cell if it is a mobile number.

Moreover, the analysis of the large amount of data that is nowadays generated from our eveyday use of mobile phones plays a crucial role in modern criminal intelligence. By leveraging graph analytics, network clustering, and statistical methods, (Ferrara et al. 2014) systematically show how it is possible to map and analyze criminal organizations based on phone communication data.

> *"Social Network Analysis has proved a valuable tool in order to ascertain the central members of criminal networks, the existence of subgroups, the interactions among individuals and subgroups, the flow of information in the network, the sensitive members and/*

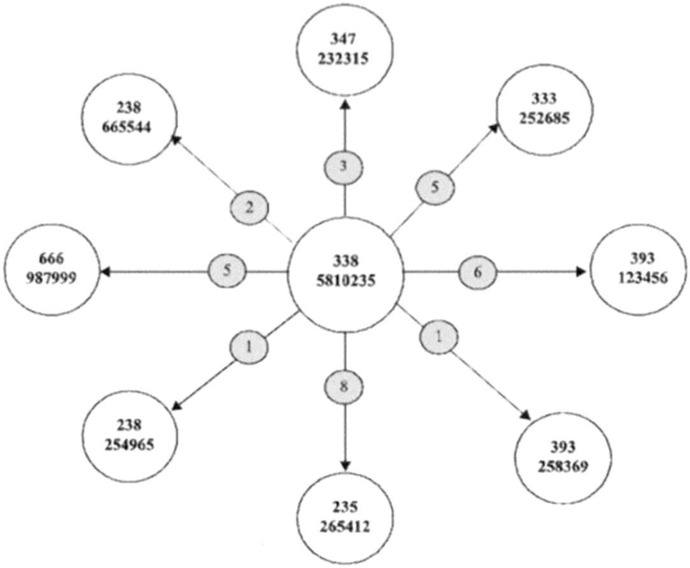

Fig. 11 Representation of a telephone flow

> *or relations whose removal could eventually lead to the destruction of the network. In this context, the analysis of phone call networks is crucial to gain fundamental information about interconnectivity and communication among criminals, and to progress fruitfully the investigations"* (Ferrara, De Meo, Catanese, and Fiumara, 2014).

This reinforces the modern significance of telephone flow analysis as a strategic tool for dismantling criminal organizations by identifying critical nodes in their communication infrastructure.

5 Conclusions and Current Prospects

Nowadays, the integration of Artificial Intelligence and machine learning into the much broader social network analysis systems to pursue criminal intelligence analysis has significantly enhanced how illicit networks can be analyzed.

Research by (Lopes et al. 2022), for example, highlights how graph representation learning and machine learning methods can predict missing criminal partnerships with remarkable accuracy, even anticipating future criminal associations.

> *Here, our goal is to fill this gap by presenting a comprehensive investigation of political corruption, criminal police intelligence, and criminal financial networks. We rely on the node2vec32 method for obtaining vector representations of nodes and edges from these criminal networks, which are then combined with simple machine learning methods in a series of predictive tasks. Our results demonstrate that network properties extracted from node2vec are effective in predicting randomly-removed partnerships from criminal networks*

and recovering missing relationships with accuracy as high as 98%.(Lopes, Requiao Da Cunha, Martins, Gonçalves, Lenzi, Hanley, Perc,and Ribeiro, 2022).

Furthermore, this integration of AI into traditional intelligence processes allows for more sophisticated criminal profiling, particularly in areas such as money laundering detection and organized crime disruption, with the aim of enhancing traditional cyber intelligence analysis frameworks (Martino 2024; Luque et al. 2023; Payá and Delgado 2017).

As the complexity of criminal enterprises continues to evolve with the development of new technologies and domains where crimes can take place, the future of criminal intelligence analysis will most likely rely heavily on the synergy between AI, machine learning, and traditional investigative techniques such as link and social network analyses (Mazurier et al. 2019; Luque et al. 2024; Sanz et al. 2024).

References

Ben-Israel, I. (1989). Philosophy and methodology of intelligence: The logic of estimate process. Intelligence and National Security, 4(4), 660–718. https://doi.org/10.1080/02684528908432023

Caligiuri, M., & Pili, G. (2021). Italian intelligence studies literature: Understanding the state of play. A comparative perspective. The International *Journal of Intelligence, Security,*

Cannavicci, M. (2006). Investigative intelligence. Giornale di Medicina Militare, 156(4), 461–464.

Chapanis, A. (1959). Research techniques in human engineering. Baltimore: Johns Hopkins Press.

Contreras, V. (2026). War without accountability. AI and the fragmentation of responsibility in contemporary warfare: Power infrastructures and technological sovereignty. Cuadernos de RES PUBLICA en derecho y criminología, 1–32. https://doi.org/10.46661/respublica.13315

International Association of Chiefs of Police. (2021). Criminal intelligence. In IACP.

Delgado Moran, J. J, Mazurier, P.A. & Paya Santos, C. A. (2019). The race to securitize the arctic in a post-cold War scenario. *Revista de Pensamiento Estratégico y Seguridad CISDE*, 4(1), 59–64. http://hdl.handle.net/10272/17180

Di Paolo, A. M. (2000). Elementi di intelligence e tecniche di analisi investigativa. Rome: Laurus Robuffo.

Dintino, J., & Martens, F. T. (1981). Enduring enigma: The concept of intelligence analysis. Police Chief, 48(12), 58–64.

F. B. Gilbreth & L. M. Gilbreth. (1917). The MacMillan Company. New York, pp. 1–30.

Fernández Rodríguez, J.C & Liz Rivas, L. (2023). El terrorismo: aspectos psicosociales en el proceso de radicalización. en: "Repercusiones de la radicalización yihadista en la seguridad Europea, Mediterránea y Latinoamericana", Thomson Reuters Aranzadi. Pp. 270–287. https://doi.org/10.5281/zenodo.14540821

Ferrara, E., De Meo, P., Catanese, S., & Fiumara, G. (2014). Detecting criminal organizations in mobile phone networks. Expert Systems with Applications. https://doi.org/10.1016/j.eswa.2014.03.024

Gangi-Guillén, G. K. (2025). Derechos humanos y derecho penal en la era de la inteligencia artificial: retos y propuestas. Cuadernos de RES PUBLICA en derecho y criminología. https://doi.org/10.46661/respublica.11635

Godfrey, E. D., & Harris, R. (1971). Basic elements of intelligence. Washington, D.C.: U.S. Government Printing Office.

González Trigo, N. A. (2023). El agente encubierto ante la criminalidad organizada transnacional. *Cuadernos de RES PUBLICA en derecho y criminología,* (1) 85–94. https://doi.org/10.46661/respublica.8062.

Harper, W. R., & Harris, D. H. (1975). The application of link analysis to police intelligence. Human Factors, 17(2), 157–164. https://doi.org/10.1177/001872087501700206

Harris, D. H. (1978). Development of a computer-based program for criminal intelligence analysis. Human Factors: The Journal of the Human Factors and Ergonomics Society, 20(1), 7–14. https://doi.org/10.1177/001872087802000107

Harris, D. H. (1978). Link analysis of threats and physical safeguards. In J. J. Kramer (Ed.), The role of behavioral science in physical security: Proceedings of the third annual symposium (pp. 17–22). Washington, D.C.: National Bureau of Standards.

Ianni, F. A. J., & Reuss-Ianni, E. (1990). Network analysis. In P. P. Andrews & M. B. Peterson (Eds.), Criminal intelligence analysis (pp. 67–84). Loomis, CA: Palmer Enterprises.

International Association of Chiefs of Police, Criminal Intelligence, in IACP (2021).

Izzi, S. (2011). Intelligence e gestione delle informazioni. L'attività preventiva contro i traffici illeciti. Milan: Franco Angeli, pp. 23–24.

Johnston, R. (2003). Developing a taxonomy of intelligence analysis variables. Studies in Intelligence, 47(3), 61–71.

Lippert, S. (1971). Travel in nursing units. Human Factors, 13(3), 269–282. https://doi.org/10.1177/001872087101300308 PMid:5558220

Lopes, D., Requiao Da Cunha, B., Martins, A., Gonçalves, S., Lenzi, E., Hanley, Q., Perc, M., & Valentin Ribeiro, H. (2022). Machine learning partners in criminal networks. Scientific Reports. https://doi.org/10.1038/s41598-022-20025-w

Lupsha, P. A. (1980). Steps toward a strategic analysis of organized crime. The Police Chief, 47(4), 36–38.

Luque Juárez, J. M., Payá Santos, C. A., & Arenas Morales, F. (2023). Contexto de las políticas de seguridad ciudadana. *Cuadernos de RES PUBLICA en derecho y criminología*, (2), 69–82. https://doi.org/10.46661/respublica.8293

Luque Juárez, J. M., Sanz González, R., Payá Santos, C., Rodríguez González, V. (2024). Aplicaciones de la inteligencia artificial en la criminología y ciencias policiales. https://doi.org/10.13140/RG.2.2.11118.63041.

Martens, F. T. (1987). The intelligence function. In H. Edelhertz (Ed.), Major issues in organized crime control (p. 133). Washington, D.C.: U.S. Department of Justice, National Institute of Justice.

Martino, L. & Merenda, F. (2021). Artificial intelligence: A paradigm shift in international law and politics? Autonomous weapon systems as a case study. in: Giampiero Giacomello & Francesco N. Moro & Marco Valigi (ed.), Technology and International Relations, chapter 5, pages 89–107, Edward Elgar Publishing. https://doi.org/10.4337/9781788976077.00012

Martino, L. (2024). Cybersecurity in Italy. Governance, Policies and Ecosystem. Springer Nature. https://doi.org/10.1007/978-3-031-64396-5

Mazurier, P, A., Delgado Morán, J, J. & Paya Santos, C, A. (2019). Gobernanza constructivista de la internet. *Teoría y Praxis*, 17(34), 107–130. https://doi.org/10.5377/typ.v1i34.14823

McDowell, D., Wardlaw, G., & Schmidt, J. (1991). A model for the development of strategic crime assessments. Canberra: Attorney-General's Department of Australia, pp. 11, 22.

Mena, J. (2003). Investigative data mining for security and criminal detection. Amsterdam: Butterworth-Heinemann, p. 80.

Otte, E., & Rousseau, R. (2002). Social network analysis: A powerful strategy, also for the information sciences. Journal of Information Science, 28(6), 441–453. https://doi.org/10.1177/016555150202800601

Payá Santos, C., & Delgado Morán, J. J. (2017). Uncertainty of dimensional analysis of intelligence. *URVIO Revista Latinoamericana de Estudios de Seguridad* 21, 225–239.https://doi.org/10.17141/urvio.21.2017.2962

Payá-Santos, C. A. (2023). El desempeño de la inteligencia en España en el ámbito público, empresarial y académico. *Revista Científica General José María Córdova, 21*(44), 1029–1047. https://doi.org/10.21830/19006586.1222

Sanz González, R, Luque Juárez, J M.ª, Martino, L, Liz Rivas, L, Delgado Morán, J. J, & Payá Santos, C. A. (2024) Artificial Intelligence Applications for Criminology and Police Sciences. *International Journal of Humanities and Social Science.* Vol. 14, No. 2, pp. 139–148. https://doi.org/10.15640/jehd.v14n2a14

Schneider, R. S. (1995). The criminal intelligence function: Toward a comprehensive and normative model. IALEIA Journal, 9(2), 403–424.

Sparrow, M. K. (1991). Network vulnerabilities and strategic intelligence in law enforcement. International Journal of Intelligence and CounterIntelligence, 5(3), 255–279. https://doi.org/10.1080/08850609108435181

Steele, R. D. (2000). On intelligence: Spies and secrecy in an open world. Oakton, VA: OSS International Press.

Stephen, R., & Schneider, R. S. (1995). The criminal intelligence function: Toward a comprehensive and normative model. IALEIA Journal, 9(2), 403.

Stockley, D. (1988). National Drugs Intelligence Unit. Police Journal, p. 299. https://doi.org/10.1177/0032258X8806100402

Stuster, J., & Chovil, A. (1994). Tracking disease outbreak. Ergonomics in Design: The Quarterly of Human Factors Applications, 2(3), 8–15. https://doi.org/10.1177/106480469400200304

Thomson, R. M. (1972). Design of multi-man-machine work areas. In H. P. Van Cott & R. G. Kinkade (Eds.), Human engineering guide to equipment design (pp. 133–144). Washington, D.C.:

U.S. Government Printing Office.

United Nations Office on Drugs and Crime. (2011). Criminal intelligence manual for analysts. Vienna: United Nations Office on Drugs and Crime.

Xu, J. J., & Chen, H. (2004). Fighting organized crimes: Using shortest-path algorithms to identify associations in criminal networks. Decision Support Systems, 38, 473–487. https://doi.org/10.1016/S0167-9236(03)00117-9

Artificial Intelligence Applied to Criminology and Criminal Science. Challenges of Penal Responsibility

Lenny Liz-Rivas and Juan-José Delgado-Morán

Abstract The concept of artificial intelligence first emerged at the beginning of the twentieth century with the goal of enabling machines to perform tasks similar to those performed by humans. Since then, various theories have been presented regarding the scope of liability for errors made by artificial intelligence systems, particularly from a criminal law perspective. This study aims to outline this issue through an examination of the legal hurdles surrounding the concept of penal liability for actions taken by artificial intelligence. Furthermore, the objective of this study is to analyse the ethical and social implications arising from the adoption of artificial intelligence in penal justice and provide recommendations for the responsible development and use of artificial intelligence technologies in penal justice, identifying the key factors affecting the effectiveness and fairness of artificial intelligence tools in the penal justice sphere. The aim of the study is to demonstrate the applications of artificial intelligence, identify its corresponding areas, clarify its benefits and predict its anticipated impact. Furthermore, it seeks to highlight proposed penal law regulations governing the use of artificial intelligence and analyse the suitable sanctions suggested for artificial intelligence entities.

1 Introduction

With the emergence of artificial intelligence, society stands at the threshold of a digital revolution that is expected to significantly transform the way people live. In light of these advancements, it is crucial to update and establish laws and regulations that align with this new reality (Delgado et al. 2019). The fourth industrial revolution has brought about significant technological advancements such as artificial intelligence,

L. Liz-Rivas
Department of Law and International Relations, Nebrija University, Madrid, Spain
e-mail: llizri@nebrija.es

J.-J. Delgado-Morán (✉)
Department of Law, Pablo de Olavide University, Sevilla, Spain
e-mail: jjdelmor@upo.es

L. A. Garcia-Segura (ed.), *Collaborative Insights on Technology and Legal Frameworks*, Advanced Sciences and Technologies for Security Applications,
https://doi.org/10.1007/978-3-032-27714-5_8

robotics, the Internet of Things, autonomous vehicles, 3D printing, nanotechnology, biotechnology, materials science, government informatics, and Blockchain, among others, and it is essential to establish a comprehensive legal framework to regulate these new realities (Payá et al. 2025c; Mazurier et al., 2019). It is essential to have a realistic understanding of the applications of artificial intelligence, its impact on human and social behavior, and the legal interests that need to be safeguarded (Payá et al. 2023).

2 Characteristics and Domains of Artificial Intelligence

Artificial intelligence systems possess various distinctive characteristics, with their most notable trait being their ability to learn, make logical deductions, and respond to unforeseen situations that are not explicitly defined within the system. Applications of artificial intelligence have expanded into various sectors, including commerce, the economy, and industry. The use of robots in industry or the healthcare sector, such as surgical robots performing complex surgeries, demonstrates the growing application of artificial intelligence in complex processes (Payá et al. 2025a; 2025b). It is therefore corroborated that artificial intelligence will have a significant impact on various aspects of human life, including education, transportation, communications, traffic, the environment, security, defence, and social media (Hao 2022; Herrera 2024).

Three distinct types of artificial intelligence are suggested: specialised artificial intelligence, general artificial intelligence, and super artificial intelligence. Specialized artificial intelligence refers to artificial intelligence systems that can perform specific tasks, such as managing autonomous vehicles, identifying speech or images, or less or more complex recreational tasks in intelligent devices (Martino and Merenda 2021; Mazurier et al. 2019). This type of artificial intelligence is very common and accessible to everyone in the present day. Artificial general intelligence encompasses a form of AI capable of executing tasks at a level equivalent to human capabilities, as noted in a 2019 study by Mazurier et al. Regarding cognition, its focus is on enabling machines to think and plan autonomously, much like human cognition (Payá and Delgado 2025). Artificial superintelligence, with a supposed potential making it superior to human intelligence, possesses the ability to surpass individuals with specialisation and knowledge in various tasks (Payá et al. 2025c).

This type of artificial intelligence should have several key characteristics, such as learning, planning, autonomous communication, and the ability to make informed decisions, with all aspects of life, including applied science, aiming to provide artificial intelligence with skills similar to human intelligence in order to carry out any type of operation (Martino and Merenda 2021; Hao 2022). Although the concept of superior artificial intelligence is a theoretical idea, it is currently believed not to exist. Despite the advancements in artificial intelligence, it is progressing towards automatic learning and deep learning, resulting in Generative Artificial Intelligence (Payá et al. 2025a; 2025b; 2025c; Sanz et al. 2024).

3 The Advantages and Disadvantages of Artificial Intelligence

Despite the beneficial characteristics of artificial intelligence, various projections suggest it also has a negative aspect, making it a conducive environment for the emergence of new forms of offences. This has led for example, to Europol (2019), emphasising the need to anticipate the repercussions of emerging technologies that could be utilised by crime. Researchers have highlighted the potential of artificial intelligence to provide new methods for criminals, hackers, and attackers to carry out malicious activities (Fernández y Sanz 2021; Payá et al. 2025a; 2025b; 2025c). This encompasses the use of artificial intelligence to amplify the effectiveness of phishing attacks through the creation of fraudulent emails and unfavourable responses, or the widespread manipulation of visual or auditory content, known as deepfakes, used to carry out massive disinformation campaigns, allowing them to conduct advanced and dangerous attacks (Sanz et al. 2024).

Based on this knowledge, they will have the ability to make decisions autonomously, replicating or independently utilizing any known security vulnerability to gain an advantage over interconnected computer systems in a network (Martino and Merenda, 2021). According to Europol (2019), law enforcement agencies need to rapidly enhance innovation, creativity, dynamism, and the modernization of their protocols to effectively counter criminal threats by leveraging new technologies, utilising more effective methods for detecting and preventing crime, allocating resources in a more suitable manner, and improving decision-making processes (Payá and Delgado 2025; Sanz et al. 2024).

The importance of addressing these specific forms of artificial intelligence highlights the potential that artificial intelligence applications could achieve in the immediate future and the level at which the principles of penal law are considered applicable, motivated by the need to have more effective methods for detecting and preventing the negligent or criminal use of these, by allocating resources in a more suitable manner and improving decision-making processes (Payá and Delgado 2025). Although artificial intelligence can enhance efficiency, precision, and equity, it is crucial to guarantee that these technologies are used in a manner that respects the rights, dignity, and autonomy of individuals (Samoili et al. 2020).

4 Normative Responsibility of Artificial Intelligence

In the absence of regulation, there has arisen an urgent need for an international framework to govern the uses of artificial intelligence, taking into account regional and international movements to address the existing void. Notwithstanding these advances, they also bring intense legal debate regarding the need to implement a legal regulation for the use of artificial intelligence, which is divided into two positions: the first position argues that the scope of innovation and progress in the field of artificial intelligence should not be limited by legal regulation (Herrera 2024).While a sector of legal professionals considers it essential to have a legal framework governing the use of artificial intelligence across all domains, as well as establish an ethical framework to oversee its use and prevent its misuse, which could harm people. The rapid advancement of artificial intelligence technology has also posed significant challenges in determining criminal liability when artificial intelligence systems cause harm (Hao 2022; Herrera 2024).

The importance of having a legal framework that regulates its uses highlights existing concern about the increasing adoption of artificial intelligence technologies and their availability for treating personal data which it may infringe upon in personal spheres. Furthermore, the expansion of these uses has led to the emergence of new actors in the area of legal liability norms, whether penal, civil or administrative, which implies examining the responsibility of programmers, manufacturers, developers, operators, users and owners of these technologies, as well as the legal boundaries of their liability for damages that may result from their use (Herrera 2024).

The European Union stood out in this area, and these issues were of great significance to its experts, given that the European Parliament's resolution (2021) on the use of artificial intelligence in the context of penal law and the application of the law stressed the need for a clear legal framework to establish liabilities resulting from its use. Criminal liability for the potentially damaging effects of artificial intelligence systems is established as long as regulations governing this liability maintain personal culpability for harms resulting from these applications. The efforts of the European Union ultimately led to the implementation of the first international treaty to govern this issue, Regulation (EU) 2024/1689.

The function of legal scholars is of paramount importance when crafting regulations, guaranteeing they are equitable and functioning properly. Existing legal frameworks typically fail to account for the intricacies of artificial intelligence, thereby necessitating the creation of novel regulatory approaches. This study investigates the feasibility and obstacles associated with implementing proposed legal frameworks for the criminal accountability of artificial intelligence systems. As artificial intelligence continues to advance and become more pervasive, the future of criminal law must be properly prepared and adapted to effectively address the complex issues and nuances related to artificial intelligence's penal responsibility. This entails conducting a thorough analysis of existing legal frameworks and developing updated regulations that take into account the unique characteristics of artificial intelligence systems and their potential consequences in terms of penal responsibility. In this way, society can ensure that the legal system is equipped to handle any problems that arise from artificial intelligence technology (Herrera 2024).

5 The Legal Regulation of Artificial Intelligence in the European Union

The Council of Europe has approved the first international treaty on artificial intelligence, Human Rights, Democracy, and the Rule of Law (2024). The level of interest the European Union attaches to artificial intelligence specifically, and the need to establish a cohesive framework to regulate it, anticipating the expected evolution in its use, is thus as evident as a reality (Samoili et al. 2020). These legislative efforts represent an innovative model that undoubtedly will help promote the innovation and development of artificial intelligence from a clear legal perspective that highlights the legal boundaries in the use of this rapidly advancing technology (Samoili et al. 2020). The agreement consisted of 36 points. The treaty outlines a legal framework based on risk assessment (risk management and impact framework), and article (1, Sect. 1) of the treaty states that its purpose is to ensure complete compatibility among all activities forming part of the complete life cycle of an artificial intelligence system (industry, programming, application, and operation). Additionally, human rights, democracy, and the rule of law must ensure that risks, such as job loss and privacy violations, arising at each stage are regulated to mitigate the negative overall effects, such as social inequality and personal autonomy loss, on fundamental rights and the rule of law, without compromising the need to stimulate creativity; however, it is essential that this creativity be responsible and not develop without any type of control or connection (Regi 2026).

The second chapter of the treaty consists of two articles that lay down general obligations for member states with regard to implementing necessary measures to ensure that artificial intelligence systems comply with human rights (Article 4). It is also emphasized that artificial intelligence systems should not be used to compromise the integrity, independence, and effectiveness of institutions and the democratic process, as well as the importance of respecting the judicial power's sovereignty, access to justice, and the ability of individuals to form their opinions freely.

This chapter also covers a set of obligations related to the most significant infringements arising from artificial intelligence applications and technologies, [number not provided]. For instance, article (7) states that measures must be implemented to ensure respect for human dignity and individual autonomy, whereas article (8) refers to the adoption of all necessary measures to protect human dignity and individual autonomy. A member state takes the necessary steps to consider transparency and oversight. According to article (9), measures must be taken to ensure accountability and responsibility, whereas article (10) stipulates that necessary steps must be taken to ensure equality and non-discrimination. On its part, article (11) suggests that measures must be implemented to safeguard personal privacy and data. The article (12) states that necessary measures must be taken to ensure the reliability of artificial intelligence systems. The article (13) refers to the implementation of necessary actions to ensure safe innovation.

The European Parliament has worked to establish the 'human responsible agent theory' with the aim of regulating European civil law standards related to robots, which was published by the European Union's Legal Affairs Committee in February 2017. The European Robotics Law acknowledges that it is impossible to assign responsibility to a robot, understood as an automated entity lacking fundamental human logic and the ability to think about potential harm to a third party. Therefore, a human representative is accountable for the robot's actions or inactions. This originates from the human being being perceived as a matter of technological emulation (Samoili et al. 2020).

With the emergence of new generations of robots, the European Union foresees an electronic legal personality for these, granting them rights and establishing obligations, considering artificial intelligence systems and robots as objects, with their owner viewed as a custodian of these objects, subject to possible error, just like a conventional car owner (Vásquez 2022). The EU's plans for liability in relation to artificial intelligence are insufficient in terms of specific regulations on the development and implementation of artificial intelligence, and the document suggests revisions to improve the proposed framework (Herrera 2024).

6 The Penal Liability of Artificial Intelligence

The primary aim of penal law is to protect individuals, their rights, and possessions by means of regulations that define the criminal liability of those who break the law and threaten legally protected interests, and it does this through the legally established penalties that act as deterrents to offenders. It is therefore also the task of the Criminal Code to establish the criminal liability of those who violate its provisions and detail the specific penalties that correspond to them. Consequently, criminal liability is the foundation of the penal justice system, establishing the conditions under which individuals can be held accountable for actions that violate the law (Fernández y Sanz 2021). Criminal liability involves a person's legal accountability for committing a crime, which can lead to various penalties. Penal liability is based on principles established in constitutional laws and in Criminal Law. A crucial aspect of penal liability. For a person to be considered legally responsible, they must have acted with the intent to commit the crime or been aware of the illegality of their actions, distinguishing between deliberate and involuntary actions, which significantly impacts the evaluation of the proportionality of their punishment (Hao 2022; Herrera 2024).

To establish criminal liability, certain requirements must be met, with the first being that the perpetrator is a physical or legal person, as criminal liability does not arise from the actions of a machine or an animal (Villalobos 2022). In the second place, it is crucial that this individual or entity be attributed with an act that is classified as a crime under the Penal Code, and the third element is that there be imputability in the person who committed the crime. Artificial intelligence applications aim to train machines to perform tasks similar to those of humans, which can result in errors that could potentially be considered crimes under the Penal Code. This creates various legal complexities due to lawmakers' difficulty in defining the legal personality of artificial intelligence systems, and currently, there are differing debates in jurisprudence regarding the possibility of granting legal personality to these applications, and discussions about the nature and scope of criminal liability that can arise from errors committed by artificial intelligence applications (Hao 2022; Herrera 2024).

With the advent of artificial intelligence and its applications in various aspects of life, penal jurisprudence initiated a debate on the penal responsibility related to artificial intelligence, which can be classified into three categories. The first position holds that artificial intelligence applications have no penal responsibility, the second seeks new legal ideas that include the possibility of penal responsibility for artificial intelligence applications, and the third endeavours to harmonize the application of traditional norms to artificial intelligence applications (Hao 2022; Herrera 2024).

Those who defend traditional penal jurisprudence believe that it is inconceivable to hold artificial intelligence applications criminally responsible. These applications lack human characteristics, and criminal liability can only be attributed to humans, not animals or machines (Fernández y Sanz 2021). Artificial intelligence applications are essentially machines that cannot be conceived with criminal liability, as they are not human beings (Herrera 2024). The essence of robots and other artificial intelligence applications precludes assigning them a crime, since the foundation of penal responsibility is based on the freedom of choice, which robots and artificial intelligence applications do not possess (Vásquez 2022).

Therefore, if the author lacks the freedom to choose, there is no reason to hold them accountable (Villalobos 2022; Herrera 2024). Therefore, criminal behavior cannot be attributed to a robot lacking freedom of will and choice. The legal standard primarily focuses on physical individuals who are capable of understanding and assuming responsibility for violating the standard (Vásquez 2022). Intentional criminal behavior is a deliberate action, and the will constitutes the moral essence of the crime, applying solely to human beings. The individual being held accountable for the crime must possess robust mental and intellectual abilities and be aware of the action taken (Herrera 2024).

On the other hand, trends exist to establish a theoretical framework that manages the penal liability of artificial intelligence entities independently of conventional norms of penal law. Proponents of this trend question whether it is necessary to establish an autonomous penal law for artificial intelligence, emphasizing the need to establish arguments and discussions about the necessity of an appropriate penal law for artificial intelligence (Hao 2022; Herrera 2024). According to these trends, having an autonomous penal code would be advantageous to society, given that technology advances at an accelerated pace, and its evolution occurs more rapidly than the legislator can adapt):

7 Models for the Penal Responsibility of Artificial Intelligence

This model provides a basic framework, disregarding human attributes during the creation of artificial intelligence. In this model, artificial intelligence is viewed as an innocent agent. Notwithstanding, this model disregards the potential of artificial intelligence to commit crimes. In artificial intelligence, the tool functions as a mediator, while the perpetrator is the original actor. Two individuals can assume the role of the original actor in artificial intelligence: the programmer and the end-user of artificial intelligence (Hao 2022).

Notwithstanding, the usefulness of this model would decline when artificial intelligence chooses to act entirely independently, although it was not designed for this purpose. This model establishes the criminal liability of the programmer or the artificial intelligence end-user, even if the AI was not designed or used to commit the crime, as long as the crime occurs as a natural and predictable consequence of the robot's artificial intelligence behavior (Vásquez 2022). Therefore, this model does not require human actors' intentions; only their lack of care falls within the scope of criminal liability (Herrera 2024). This restriction highlights the complex and crucial task of determining the criminal liability of artificial intelligence entities, presenting a significant intellectual challenge for the field of criminal law.

For an artificial intelligence entity to be held criminally liable, it is required that the same conditions as in the individual be met, namely that it acted intentionally or with knowledge. The primary focus here is not the action itself. When artificial intelligence manages its mechanisms and, consequently, acts autonomously, the requirements for action can be deemed satisfied. Although an artificial intelligence entity possesses a level of creativity, environmental understanding, and algorithms that, in certain cases, simulate the human brain's thought process, it lacks emotions (Hao 2022).

In many cases, even among individuals, emotions are often irrelevant, as certain crimes solely involve a component of knowledge. From this viewpoint, it logically follows that even an entity of artificial intelligence could commit a crime. The theory behind this model has been expanded to take into account the relationship between humans and artificial intelligence entities as collusion and collaboration, enabling the corresponding criminal liability to be assigned (Herrera 2024). If we move away in time from the conventional definition that a crime is a deliberate human act, this third model could be of some use. Despite this, it is essential to adjust certain fundamental principles of penal law to incorporate the penal responsibility of artificial intelligence entities into various jurisdictions, addressing the responsibility of artificial intelligence in relation to human users, programmers, owners, and the manufacturing company, as stated below (Herrera 2024).

It is essential to differentiate between various forms of artificial intelligence when holding systems of artificial intelligence accountable for their actions through legal liability. Specialized artificial intelligence refers to systems capable of executing specific and clearly defined tasks, such as autonomous vehicles (Martino and Merenda 2021). On the other hand, artificial intelligence seeks to mimic human thought and planning abilities, enabling machines to think and plan independently, and if they provide incorrect or misinterpreted responses, actions, or information, individuals are responsible for their decisions based on it (Herrera 2024).

The question is whether artificial intelligence can be seen as an intermediary in the commission of a crime, drawing parallels with the theory of moral agency, which is applied when an immature individual or an animal commits an illegal act. As a result, the artificial intelligence program could be seen as an impartial intermediary. In this scenario, it is assumed that the artificial intelligence entity lacks cognitive abilities (Herrera 2024). Therefore, there is no legal distinction between the machine used in a crime and artificial intelligence. In this context, the underlying assumption is that the artificial intelligence entity does not rely on a specific programmer or user. Therefore, in order for an artificial intelligence entity to be held accountable for a crime, three conditions must be met.

- It is essential to establish that the crime occurred as a direct result of the actions of the artificial intelligence entity itself.
- For the entity of artificial intelligence to be held accountable, it must be proven that it is responsible for the crime independently, without relying on a programmer or user.
- It must be proven that the entity was involved in the criminal act with knowledge or intent.

Let's assume that the artificial intelligence program or system controls the movement of the machine's movable parts, whether mechanical or of another type. Consequently, any action, whether beneficial or detrimental, undertaken by this machine can be attributed to the artificial intelligence entity, taking into account the particular characteristics of artificial intelligence methods (Gangi 2025). In general, we can conclude that the legal liability for crimes committed by artificial intelligence involves discussing the culpability of creators and producers of artificial intelligence systems, as well as users of artificial intelligence (Hao 2022). Future developments in penal liability laws may align with advancements made in this area. In this context, we examine various hypotheses that govern the process of attributing liability to individuals for crimes committed due to the misuse of artificial intelligence systems, in accordance with the principles established by penal law. This is because the person may not be directly accountable for the crime. Punishment is only imposed on an individual if he is deemed guilty or an accomplice (Herrera 2024).

8 Primary Responsibility of the User as the Original Perpetrator of the Crime

To examine the liability of the owner or user, we need to focus on three hypotheses: The first hypothesis involves the use of artificial intelligence to commit crimes. In this hypothesis, it is assumed that the perpetrator of the crime uses an intelligent machine to carry out criminal activities (Villalobos 2022). To establish criminal liability, it is crucial to establish a causal link between the perpetrator's actions and the criminal outcome achieved through the use of artificial intelligence. This would imply that the offender committed the crimes knowingly, with the machine's assistance. The foundation of artificial intelligence systems is that the penal legislator has at their disposal the same tools for committing a crime as the perpetrator, who can use a stone or a gun to attack the victim, just as an animal can be used for the same purpose or a artificially intelligent computer can be manipulated to achieve a similar outcome (Villalobos 2022). In the field of law, it is crucial to provide evidence showing that the artificial intelligence system was under the control of the perpetrator, that the perpetrator had prior knowledge of its use, and that they directed their actions with the intention of committing the crime using the artificial intelligence's capabilities (Villalobos 2022). Notwithstanding, if the artificial intelligence entity deviates from the original plan and commits an additional or entirely different crime, in this context, it can be argued that the programmer could be considered criminally liable intentionally.

The second hypothesis proposes the concept of probabilistic intentionality. The hypothesis proposes that if standard procedures of an artificial intelligence system are misused to commit a crime, the system itself could be held accountable for the offense. A previous view suggests that the machine's programmer should be considered liable if they knew there was a possibility of this illegal outcome occurring prior to its occurrence. In this context, the person in charge took a specific action, anticipating that it would lead to a particular outcome.

The third theory suggests that negligence can occur during the manufacturing, programming, or use of artificial intelligence. In this instance, it is assumed that the individual involved in the crime, whether a user, a manufacturer or an artificial intelligence programmer, did not intend to use it for criminal purposes. Despite this, the crime occurred due to her negligence, specifically because she failed to follow the care and caution guidelines. In this case, the individual who intentionally commits the illegal act, whether it be a user, manufacturer, or programmer, will be held criminally liable. It is essential to establish a link between the commission of the crime and the involuntary mistake made by the perpetrator (Villalobos 2022).This occurs when the machine code fails to anticipate all possible scenarios or when the crime is unforeseen. The responsibility lies with individuals who have committed a crime when the owner improperly uses it for criminal purposes. The liability does not automatically transfer from the machine to the human, and this person must acknowledge their responsibility for any crime committed by these systems or applications (Herrera 2024).

In contrast to this viewpoint, it is argued that the penal responsibility attributed to the errors of artificial intelligence should be replaced by the responsibility based on risk assumption. It is argued by the previous perspective that the owner or user should be held accountable for the crimes committed through the artificial intelligence under their control, based on the concept of assumed responsibility. Therefore, it is her responsibility to show a "lack of accountability". Some proponents of the previous viewpoint argue that liability can be attributed under the natural possibility responsibility model. According to this model, the programmer or user can be held liable if the crime committed was a predictable and probable result of the entity of artificial intelligence's behavior.

The opinion pertains to the ability of programmers or users to foresee potential irregularities by an artificial intelligence entity. Therefore, if the crime committed by the artificial intelligence is a direct and predictable result of the actions of the programmer or the user, they can be held accountable. This responsibility is based on realistic possibilities and does not require additional factors beyond programmer or user negligence. It is not necessary for programmers or users to have any knowledge about the commission (Herrera 2024).

9 Regulation of Artificial Intelligence in Sexual Crimes Within Spanish Legislation

The increasing use of artificial intelligence presents significant opportunities which provide substantial benefits, efficiency, accuracy, and appropriateness, ultimately leading to a transformation that is estimated to be positive for society and the economy as a whole, although it also generates extraordinary risks (Herrera 2024).

The development of artificial intelligence is fostering the emergence of new types of criminal activity, with severe consequences for the mental health of the victims (Liz 2023; 2024). As a foundational element and starting point, artificial intelligence cannot infringe on any fundamental right or freedom, and must be worthy of public trust, working in the service of people, with AI systems designed to protect and benefit all members of society (Samoili et al. 2020). According to the European Commission (2020), artificial intelligence is defined as software systems (and possibly also hardware systems) designed by humans that operate in the physical or digital dimension in response to a complex objective.

- Perceiving their surroundings through the acquisition and interpretation of both structured and unstructured data.
- Reasoning about knowledge, processing information derived from these data, and deciding the best actions to achieve the given objective.

In recent times, as a result of the widespread dissemination of pornographic images created by artificial intelligence, there have been concerns about the predictability and normative application in such situations, and more specifically, about whether an effective penal response exists. The term "Deep Learning" refers to an artificial intelligence method, for which algorithms are developed that enable a computer to learn automatically by formulating high-level abstractions in order to generate audiovisual patterns (Martino and Merenda 2021; Hao 2022).

Deepfakes of a pornographic nature represent a highly complex issue, given that the victims experience multiple violations of their personal rights, including the right to privacy, honour, and reputation, the protection of their image and voice, and the unauthorized use of personal data (Sanz et al. 2024). Using this technique, it is possible to create fake videos through automatic editing of images and sounds developed by artificial intelligence via Deep Learning (Martino 2024; Somers 2020).

At this juncture, we must consider whether there is indeed a legal remedy for such actions and sex crimes perpetrated through artificial intelligence, and specifically a penal response (Delgado and Liz 2022). As a prior consideration, we must deny the existence of any bespoke legislation in our penal law that sanctions this type of behavior. The task now is to examine under what regulatory umbrella sexual behaviors using artificial intelligence (Murcia et al. 2025) would be punishable. Additionally, before delving deeper into the subject, it's essential to highlight that on December 8th, 2023, the European Union's institutions agreed on a regulation for artificial intelligence through a Regulation, marking the first global approved standard, which will come into effect on August 1st, 2024 as previously mentioned. In this sense, Regulation (EU) 2024/1689 addresses the regulation of potential uses of artificial intelligence, identifying and limiting the risks that could arise from them. The mentioned legislation identifies the risks associated with specific uses of AI, categorizes them into four levels of risk, and establishes different regulations for each level (Herrera 2024).

Notably, behaviors that compromise an individual's image, specifically those related to artificial intelligence, would be protected under Article 7.5 of the LO 1/1982 (Gobierno de España 1982), as this article considers it illegitimate to capture, reproduce, or publish a person's image through photography, film, or any other method in private or public settings, except in cases outlined in Article 8.2. As previously noted, there is no specific legislation to penalize the commission of sexual crimes assisted by artificial intelligence, regulation of which could have been included in LO 10/2022, of 6 September, as well as in LO 4/2023 (Gobierno de España 2023).

Following the aforementioned legislative amendments, sexual offenses, as per the current regulation, are located in Title VIII, specifically relating to "Crimes against sexual liberty," articles 178 to 194 bis. Notwithstanding the lack of specific regulation on the subject, the Supreme Court has made an interpretation that would allow for the application of the mentioned criminal types to the sexual offences committed through the use of artificial intelligence, understanding that there has been an attack on online sexual freedom (Murcia et al. 2025). With this new interpretation, the traditional concept of sexual violence, understood as unwanted physical access, will gradually be left behind in favour of a broader understanding of respect for sexual freedom (Boza 2023).

In this context, the 377/2018 ruling of July 23, was the first to acknowledge the existence of sextortion as a typical modality included in online sexual offending. The conviction resulted from sexual assault charges for the introduction of bodily members or objects, and also from the crime of discovery and revelation of secrets under Article 197.1 of the Penal Code due to accessing the victims' computer and acquiring their personal data (Prosecutor's Office 2017). In this scenario, the question arose as to whether the events could constitute or not a crime against sexual liberty or indemnity (Boza 2023). The Supreme Court acknowledges the crime of sextortion in this ruling, accepting the commission of a sex-related crime online that does not require physical contact and recognising the criminal modalities specified in articles

178 and following of the Penal Code. In addition, the Supreme Court has issued verdicts on the crime of sextortion, including Sentences 311/2020 of 15 June 2020, Ref. 3777/2018, and 447/2021 of 26 May 2021, where it is acknowledged that online sexual assaults can occur due to the presence of intimidation. Noting that convictions are being handed down for the crime of sextortion, it appears challenging to apply traditional sexual offenses to those committed using artificial intelligence, as the penal codes appear to refer to actions involving real images, not images created by this type of technology, without specific judicial rulings on this matter.

Appropriately contextualizing such behaviors, would then be fitting within the crimes against intimacy and image or against moral integrity, as per traditional sexual penal codes. Therefore, although sexual crimes committed through or with the use of artificial intelligence are not specifically covered, there appears to be a jurisprudential trend marked by the Supreme Court that aims to broaden the concept of sexual freedom and its application (Murcia et al. 2025).

However, there are no specific judicial statements regarding the sexual offenses committed using artificial intelligence, and it does not seem straightforward to align them with traditional sexual offenses (Murcia et al. 2025). Artificial intelligence offers significant benefits and very positive transformation, but generates extraordinary risks. A legal response exists to counter behaviors that compromise one's image through the use of artificial intelligence, as article 7.5 of the LO 1/1982 law protects what are known as illegitimate intrusions. Within the framework of penal law, while convictions are indeed being handed down for the crime of sextortion, and there has been a gradual shift in case law aimed at recognising online sexual attacks as criminal, however, sexual crimes committed using artificial intelligence do not appear to have a straightforward place within traditional sexual offence categories, making it easier to fit them into categories such as crimes against privacy and image or against moral integrity (Murcia et al. 2025).

A legislative adaptation is proposed in this context, intended to provide specific and effective coverage for the protection of both sexual freedom and personal intimacy, which are legally protected rights affected by criminal activity. Similarly, it would be advisable to reform existing penal codes by introducing a specific mention of sexual crimes committed through artificial intelligence, adding a fifth subsection to the basic type of sexual aggression in Article 178 of the Penal Code, for purposes of typicality and with full respect for the principle of legality.

10 Challenges Regarding the Criminal Liability of Artificial Intelligence

Research suggests that the practical implementation and challenges of proposed legal frameworks for the criminal liability of artificial intelligence involve difficulties in defining accountability, establishing a legal personality for artificial intelligence, balancing the responsibility of developers, and ensuring the artificial intelligence's ethically and transparent operations (Herrera 2024; Mazurier et al. 2019). Conventional criminal law and legal theory inadequately address the liability of autonomous artificial intelligence systems due to their complexity and purported ability to make independent decisions (Martino and Merenda 2021). The criminal liability of complex artificial intelligence systems is complicated and necessitates an adaptation to the rapidly changing artificial intelligence environment, as conventional criminal law and legal theory are not equipped to address this challenge. Artificial intelligence systems can meet criteria for criminal culpability, yet the similarities and differences between human actions and those of artificial intelligence necessitate tailored regulatory strategies. The application of artificial intelligence raises ethical dilemmas and legal challenges, necessitating explicit definitions of ethical parameters and a framework for addressing violations perpetrated with artificial intelligence (Payá and Delgado 2025).

Furthermore, existing regulatory frameworks lack explicit prohibitions on crimes related to artificial intelligence, instead focusing on general legislation regarding cybercrimes and data protection. Currently, there is no regulatory framework addressing crimes related to artificial intelligence, and the issue of the legal status of artificial intelligence's criminal liability remains unresolved (Herrera 2024). Academics argue that a persistent debate exists regarding the possibility of granting legal personality to artificial intelligence, which would enable it to assume criminal and civil liability independently of its designers and users (Fernández y Sanz 2021). Furthermore, some studies indicate that the ethical parameters governing the design, development, and application of artificial intelligence require explicit definition to prevent abuses and ensure its liability Technical issues in artificial intelligence for criminal defense encompass data privacy and security, algorithm transparency, model interpretability, ethical considerations, and legal issues such as regulatory oversight of decision-making authority granted (Herrera 2024). To mitigate these risks, it is essential to develop transparent and interpretable artificial intelligence models, along with the implementation of guidelines and protective measures that ensure accountability and safeguard human rights (Fernández y Sanz 2021). When facing these challenges, artificial intelligence can make a positive contribution to improving equity and decision-making efficiency, while also adhering to the principles of justice.

On the other hand, determining liability for crimes that utilise artificial intelligence is complex, particularly distinguishing between intentional violations and negligent actionsTechnical challenges in artificial intelligence in the defence of the accused include data privacy and security, algorithmic transparency, model interpretability, ethical issues and legal challenges such as the legitimacy of evidence and regulation of decision-making (Mazurier et al. 2019). Furthermore, it is also pertinent to explicitly outline the legal obligations and responsibilities of operators, creators, and users of artificial intelligence systems. This includes potential liability for damages caused by the artificial intelligence systems developed or implemented (Herrera 2024). Moreover, the use of artificial intelligence in judicial decisions raises concerns about the possible elimination of trial and human discretion. Judicial rulings often require a nuanced understanding and evaluation of specific situations, which may not be accurately reflected by the algorithmic models of certain artificial intelligence (Fernández y Sanz 2021). Overreliance on data-driven perceptions through the use of artificial intelligence tools can harm human empathy and moral reasoning, potentially leading to decisions that are technically correct but lack compassion and fairness (Fernández y Sanz 2021). Artificial intelligence algorithms often function as "black boxes," making it difficult for law enforcement agencies and the public to comprehend how predictions are generated and decisions are made by artificial intelligence (Mazurier et al. 2019).

We should not overlook the fact that AI-related crimes often involve cross-border elements, which complicates the jurisdictional issues in applying established legal regimes and requires establishing different standards and legal frameworks to address the criminal liability of AI universally, including reviews of current legislation (Herrera 2024).

11 Conclusions

The increasing deployment of artificial intelligence applications across various fields and the resulting scenarios have raised concerns about legal liability for the consequences of using artificial intelligence. Furthermore, current regulations on criminal culpability are insufficient to establish a legal framework governing cases of misconduct related to the use of artificial intelligence. The practical challenges and implementation of the suggested legal frameworks for the penal culpability of artificial intelligence are complex. Conventional penal law is insufficient to address the complexities of artificial intelligence, which necessitates the establishment of new regulatory frameworks. The primary challenges include defining the legal status of artificial intelligence, establishing clear ethical and regulatory frameworks, determining liability, and resolving jurisdictional complexities. Proposals for new legal frameworks are seeking to establish precise rules governing the liability of artificial intelligence; nonetheless, a considerable effort is still required to ensure these frameworks are effective and comprehensive.

To maximise the benefits of artificial intelligence and minimise its risks, it is essential to adopt a holistic approach that integrates technological innovation with ethical considerations. This encompasses the development of equitable and explainable artificial intelligence models, the implementation of rigorous testing and auditing processes to identify and address biases, as well as the formulation of precise guidelines and regulations to govern the use of artificial intelligence. Through the promotion of collaboration among technologists, legal professionals, policymakers, and the community, stakeholders can ensure that artificial intelligence technologies are implemented in a way that supports justice, equity, and public trust. The focus of future research should be on discovering innovative solutions that enhance the impartiality, transparency, and accountability of artificial intelligence systems, while also examining their broader social impacts. Through these efforts, artificial intelligence has the ability to contribute positively to the transformation of the system, the improvement of results, and the promotion of principles of justice and impartiality. Ultimately, ethical concerns surrounding artificial intelligence in the realm of penal justice underscore the need to adopt a human-centered approach when designing and implementing technology, taking into account the implications of artificial intelligence on society in order to establish ethical guidelines, political structures, and governance systems that ensure the fair and equitable use of these technologies, thereby promoting justice, equity, and respect for human rights.

References

Boza Moreno, E. (2023). Delitos contra la libertad sexual: hacia una armonización Europea. *Cuadernos de RES PUBLICA en derecho y criminología*, (2) 9–16. https://doi.org/10.46661/respublica.8045

Delgado Morán, J, J, y Liz Rivas, L. (2022) Derecho penal y violencia de género en España. Algunas cuestiones a considerar. *Revista de Direito Brasileira.* Florianópolis, SC, v. 32.n. 12. p.330–343 https://doi.org/10.26668/IndexLawJournals/2358-1352/2022.v32i12.8560

Delgado Moran, J. J.; Mazurier, P.A. & Paya Santos, C. A. (2019). The race to securitize the arctic in a post-cold War scenario. *Revista de Pensamiento Estratégico y Seguridad CISDE*, 4(1), 59–64. http://hdl.handle.net/10272/17180

Fernández Bermejo, D, y Sanz Delgado, E. (2021). *Tratado de delincuencia cibernética.* Vol. 1348. Aranzadi/Civitas

FISCALÍA GENERAL DEL ESTADO (2017). Circular 3/2017, de 21 de septiembre, sobre la reforma del Código Penal operada por la Ley Orgánica 1/2015, de 30 de marzo, en relación con los delitos de descubrimiento y revelación de secretos y los delitos de daños informáticos. https://doi.org/10.62659/CF1605605

Gangi-Guillén, G. K. (2025). Derechos humanos y derecho penal en la era de la inteligencia artificial: retos y propuestas. Cuadernos de RES PUBLICA en derecho y criminología, (8), 1–14. https://doi.org/10.46661/respublica.11635

GOBIERNO DE ESPAÑA (2023). *Qué es la Inteligencia Artificial. Plan de Recuperación, Transformación y Resiliencia.* https://doi.org/10.2307/jj.17381620.5

Hao, K. (2022). What is machine learning? MIT Technology Review

Herrera De Las Heras, R. (2024) "El nuevo marco jurídico de la inteligencia artificial: las novedades del Reglamento Europeo." Actualidad civil. Editorial LA LEY. 5 6

Liz Rivas, L. (2023). La agresión sexual en los conflictos prolongados. Derecho de intervenir y obligación de proteger. Cuadernos de RES PUBLICA en derecho y criminología, (1), 71–84. https://doi.org/10.46661/respublica.8044

Liz Rivas, L. (2024).Violencia y agresión entre iguales a través de las TICS: Cyberbulling. *Alma-Mater. Cuadernos de Psicosociobiología de la Violencia: Educación y Prevención*, nº 5, 2024, Dykinson, pp. 89–105. https://doi.org/10.14679/3314

Martino, L, & Merenda, F. (2021). Artificial intelligence: A paradigm shift in international law and politics? Autonomous weapon systems as a case study. in: Giampiero Giacomello & Francesco N. Moro & Marco Valigi (ed.), Technology and International Relations, chapter 5, pages 89–107, Edward Elgar Publishing. https://doi.org/10.4337/9781788976077.00012

Martino, L. (2024). Cybersecurity in Italy. Governance, Policies and Ecosystem. Springer Nature. https://doi.org/10.1007/978-3-031-64396-5

Mazurier, P, A., Delgado Morán, J, J, & Paya Santos, C, A. (2019). Gobernanza constructivista de la internet. *Teoría y Praxis*, 17(34), 107–130. https://doi.org/10.5377/typ.v1i34.14823

Murcia Orenes, S, Salvador Peris, P y García Sánchez, Á. (2025). Delito de sexting e Inteligencia Artificial tras la Ley 10/2022, de 6 de septiembre, de garantía integral de la libertad sexual. *Cuadernos de RES PUBLICA en derecho y criminología*, (5), 01–12. https://doi.org/10.46661/respublica.11053

PARLAMENTO EUROPEO (2021). Resolución del Parlamento Europeo, de 6 de octubre de 2021, sobre la inteligencia artificial en el Derecho penal y su utilización por las autoridades policiales y judiciales en asuntos penales (2020/2016(INI))

Payá Santos, C, A, Delgado Morán, J, J, Martino, L, García Segura, L, A, Diz Casal, J, & Fernández Rodríguez, J, C. (2023). Fuzzy Logic analysis for managing Uncertain Situations. *Review of Contemporary Philosophy*. Vol 22 (1), 2023 pp. 6780–6797. https://doi.org/10.52783/rcp.1132

Payá Santos, C, A, & Delgado Morán, J, J. (2025). Cognitive biases: Understanding and mitigating their effects. American Based Research Journal. Vol (14). Issue, 4 https://doi.org/10.5281/zenodo.15412164

Payá Santos, C, A, Rodríguez González, V, Diz Casal, J, Liz Rivas, L, Sanz González, R, & Delgado Morán, J, J. (2025a). Pedagogical and technological strategies to improve the quality of teaching in the criminology classroom. *The Seybold Report.* Vol. 20, (4), pp. 252–269). https://doi.org/10.5281/zenodo.15260938

Payá Santos, C, A, Rodríguez González, V, Granados Alos, L, Liz Rivas, L, Zenaida Dominguez Pineda, N, & Fernández Rodríguez, J, C. (2025b). The impact of nonverbal communication used on TWITCH. *The Seybold Report.* Vol. 20, (04), pp. 557–574. https://doi.org/10.5281/zenodo.15332742

Payá Santos, C, A, Rodríguez González, V, Zenaida Dominguez Pineda, N, Diz Casal, J, Fernández Rodríguez, J, C, & Delgado Morán, J, J. (2025c). Role of the Human Factor in the Cybersecurity Ecosystem. *Journal of Information Systems Engineering and Management,* 10(4). https://doi.org/10.52783/jisem.v10i4.8983

Regi Rodriguez, J. (2026). Inteligencia Artificial, poder normativo y bien común en la Unión Europea. Cuadernos de RES PUBLICA en derecho y criminología, (8), 1–21. https://doi.org/10.46661/respublica.13213

Samoili S, Lopez Cobo, Gomez Gutierrez E, De Prato, G, Martinez Plumed, F; & Delipetrev B. (2020). *Samoili, Sofía Et Al. (2020). Defining Artificial Intelligence, EUR 30117 EN, Publications Office of the European Union, Luxembourg.* https://doi.org/10.2760/382730

Sanz González, R, Luque Juárez, J, M.ª, Martino, L, Liz Rivas, L, Delgado Morán, J J, & Payá Santos, C, A. (2024) Artificial Intelligence Applications for Criminology and Police Sciences. *International Journal of Humanities and Social Science.* Vol. 14, No. 2, pp. 139–148. https://doi.org/10.15640/jehd.v14n2a14

Somers, M. (2020). Deepfakes, explained. Massachusetts Institute of Technology. https://mitsloan.mit.edu/ideas-made-to-matter/deepfakes-explained

Vásquez Urosa, C. V. (2022). Las leyes de la robótica como alternativa a la regulación legal de los robots. *Revista Justicia & Derecho*, 5(1), 1–11. https://doi.org/10.32457/rjyd.v5i1.1341

Villalobos Portalés, J. (2022). La autoría de la Inteligencia Artificial en el derecho español. Revista Justicia & Derecho, 5(1), 1–19. https://doi.org/10.32457/rjyd.v5i1.1840

Artificial Intelligence and Criminal Law: New Challenges for Human Rights

Giuseppe Kodjack Gangi-Guillén

Abstract The development of artificial intelligence (AI) has transformed criminal law and the administration of justice. From crime prediction to automation of judicial decision-making, AI offers opportunities to improve the efficiency of the judicial system. However, its implementation poses serious challenges in terms of transparency, algorithmic biases, and infringement of fundamental rights. This article analyzes the impacts of AI on criminal law and in human rights, highlighting the need for an ethical and right-based regulatory framework that balances innovation and justice. A model for monitoring and auditing algorithms is proposed, as well as the use of AI as a support tool and not as the sole criteria in judicial decision-making.

Keywords Artificial intelligence · Criminal law · Human rights · Ethical regulation · Automated justice

1 Introduction

AI has allowed significant advances in crime predictive analysis and biometric identification tools. For example, it can process large volumes of data to detect criminal patterns and predict possible crimes, which allows a better allocation of police resources and the optimization of prevention strategies (Ferguson 2017). In the case of facial recognition, it has been implemented in many countries to identify suspects and solve complex cases, although its use has caused concerns for privacy and fundamental rights (Garvie et al. 2016).

Nevertheless, the use of AI in criminal law also entails significant risks such as the lack of transparency in the decision-making of the algorithms, since many of these systems work as "black boxes", it hinders the comprehension of how they come to determined conclusions (Pasquale 2015). In the judicial field, this is a challenge for the guarantee of due process, as decisions in the penal procedures must be justified

G. K. Gangi-Guillén (✉)
Universidad Antonio Nebrija, Madrid, España
e-mail: ggangi@nebrija.es

L. A. Garcia-Segura (ed.), *Collaborative Insights on Technology and Legal Frameworks*, Advanced Sciences and Technologies for Security Applications,
https://doi.org/10.1007/978-3-032-27714-5_9

and comprehensible for the citizens. Similarly, the automation of judicial decisions without adequate human supervision can lead to mistakes that violate fundamental rights such as the presumption of innocence and the right for a fair trial.

Another emerging risk is its inadequate use to commit crimes, such as generating deepfakes and digital identity theft; tools that could be used to manipulate evidence, spread misinformation or commit cyber frauds of difficult detection (Chesney and Citron 2019). The sophistication of these attacks makes the penal persecution of these crimes a challenge, due to requiring the adaptation of regulations to deal with these new crime forms. In this sense, it results crucial to establish regulatory mechanisms that balance technological development with the protection of human rights, avoiding AI turning into a breach of fundamental guarantees tool. The implementation of AI in diverse fields poses significant challenges for human rights, especially in privacy terms, due process and discrimination. One of the main risks is the violation of the privacy right, where AI technologies for facial recognition and mass vigilance allow the collection and analysis of large volumes of personal data without the appropriate consent, which could turn into an intrusive and non-authorized vigilance of individuals and communities. The Office of the United Nations High Commissioner for Human Rights (OHCHR) has signaled that the deduction and activities about tracking made by AI tools cause serious questions about privacy and data protection (The Office of the United Nations High Commissioner for Human Rights 2021).

As for due process, the opacity of many AI systems hinders the comprehension of how algorithmic decisions are made, which can lead to compromising the capacity of individuals to impugn decisions that can affect their rights, since they can't understand nor question adequately the process. Discrimination is another critical challenge associated with AI; algorithms can perpetuate or even amplify existent biases if they are trained with data that reflects social prejudices. This can result in discriminatory decisions in areas such as labor hiring, credit granting or law enforcement. An article in *Ius et Scientia* points out that one of the most important risks about automated learning is the possibility of amplifying discrimination and existent biases against certain groups.

To mitigate such risks, it's fundamental to establish regulatory frameworks that promote transparency, accountability and equity in development and implementation of AI systems. This includes the execution of evaluation of impact in human rights before the implementation of AI technologies, continuous supervision of its performance and guaranteed existence of effective mechanisms that individuals can impugn automated decisions that affect their rights.

The goal of the article is to offer a critical reflection about the incorporation of AI in the criminal field, identifying its benefits and risks from a human rights perspective. The goal is to explore how AI can contribute to the improvement of the judicial system without compromising fundamental guarantees, where solutions are proposed to balance innovation and the protection of rights. To achieve this, the main challenges about its use will be analyzed, including the lack of transparency in algorithms, the risk of algorithmic discrimination and the need to establish appropriate normative frameworks.

Likewise, strategies will be proposed to mitigate these problems, such as the creation of audits independent of AI systems, the implementation of transparency principles, the explanation of its design, and the reinforcement of the judicial supervision in automated processes.

2 The Role of AI in Criminal Justice

The integration of AI in criminal justice has generated a significant academic debate centered around its applications and both ethical and legal implications that it entails. According to Miró (2018), AI is used in police work to prevent crime through the analysis of large volume of data that allows to identify criminal patterns and predict possible criminal behavior. This predictive capability facilitates a more efficient allocation of police resources and a proactive response against crime.

However, the implementation of AI in the criminal field poses significant challenges regarding the protection of fundamental human rights. Hernández (2019) warns about the risks of algorithmic discrimination, where biases present in the data used to train the AI systems can perpetuate or even amplify existent inequalities, which in turn disproportionately affects certain social groups, it hinders transparency and accountability in judicial decision-making (Liz 2024).

In the face of these challenges, it is imperative to establish regulatory frameworks that guide the ethical and responsible use of AI in criminal justice. Balcells (2020) proposes the implementation of independent audits of algorithms and the creation of supervising organizations to ensure fairness and transparency in the application of these technologies; and also emphasizes the need for AI to act like a support tool for legal operators, without substituting the human judgment.

2.1 Vigilance and Facial Recognition

The use of AI in vigilance and facial recognition has experienced noticeable growth in the last few years, especially in the field of criminal justice. These technologies allow the identification and tracking of individuals in real time, making easier the work of security forces in prevention and solving of crimes. However, its implementation poses challenges in terms of privacy, precision and equity, which evidences the existence of benefits, but large limitations at the same time. A critical aspect is the precision of facial recognition systems, previous investigations have proved that these systems can show higher error rates in the identification of people belonging to ethnic minorities and women. For example, the study published in *Science Advances* found that the facial recognition algorithm had significantly higher error rates when identifying people with darker skin tones in comparison with individuals with lighter skin tones (Buolamwini and Gebru 2018).

Beyond concerns regarding accuracy and bias, the use of AI-driven surveillance technologies poses questions about the undermining of privacy and civil freedoms. The capacity to monitor and register movements of individuals in public and private spaces without their explicit consent can create the sensation of constant surveillance, which influences individuals' behavior and autonomy. An article published by *Big Data & Society* argues that the implementation of mass surveillance systems can lead to a normalization of state surveillance, reducing the public resistance and conscience about the implications for civil rights (Brayne 2017).

To mitigate these challenges, it is essential to establish regulatory and ethical frameworks that guide the development and use of surveillance and facial recognition technologies, the implementation of privacy impact evaluations, transparency in the algorithms employed, and the assurance of accountability mechanisms. The European union has proposed regulations that restrict the use of facial recognition systems in public spaces, emphasizing on the need to safeguard the fundamental rights of citizens (European Commission 2021), incentive that looks to balance the benefits of technology with the protection of human rights.

2.2 *Crime Prediction*

In crime prediction, AI is used to analyze large volumes of historical and recent data with the goal of identifying patterns and criminal tendencies. Through the use of automated learning algorithms, it's possible to foresee areas and moments with larger chances of criminal activity, which allows order forces to optimize the allocation of resources and the planning of more effective preventive strategies.

A recent study by Mandalapu et al. (2023) highlights that the application of profound learning techniques has improved the precision in predicting specific crimes, such as thefts and assaults, by considering variables such as geographic location, time of day and socioeconomic data.

However, the implementation of these systems is not exempt from criticism; one of the main challenges is the risk to perpetuate existing biases in the data used to train the algorithms. If the historical data reflects discriminatory police practices or social prejudices, the predictive models could reinforce these disparities, generating a disproportionate vigilance towards marginalized communities; in the words of Scantamburlo et al. (2019) the excessive dependance in the algorithmic precision could lead to the stigmatization of certain areas or groups, without dealing with the subjacent causes of the crime.

2.3 *Judicial Decision-Making*

AI has also been used to assist in risk evaluation and sentences determination. Tools like *COMPAS* in the United States analyze the factors related with criminal history and personal characteristics to influence in decisions about probation, bails and sentences, this facilitates a more objective and consistent justice administration. The investigation of Gangi (2025) has explored the use of attention models in natural language processing to predict judicial outcomes, analyzing legal texts and precedents to identify decision-making patterns.

Nevertheless, the delegation of judicial decisions to AI systems poses worries about transparency and accountability; since the algorithms used are often not transparent, making it difficult to understand how certain conclusions are reached. This can undermine the trust in the judicial system and poses questions about the equity, especially if the models incorporate implicit biases; therefore, it may reduce the consideration of mitigating or contextual circumstances that are fundamental in criminal justice (Chén 2022).

2.4 *Opportunities and Risks: Efficiency vs. Violation of Rights*

The integration of AI in the criminal field has also generated a significant debate about its efficiency and the possible violations of human rights. On the one hand, AI promises to optimize judicial processes, bettering precision in decision-making and reducing workloads on tribunals. On the other hand, worries surge about equity, transparency and the respect for fundamental procedural guarantees.

In the criminal process, it poses significant risks to human rights due to the lack of understanding in the algorithm's design, which hinders the comprehension of the criteria used in decision-making, which undermines the right to a fair trial and adequate legal representation. The dependency of historical data to train these systems can also perpetuate or even amplify biases and existent discriminations, which can disproportionately affect the processes applied to certain social groups. Regarding this, the European Parliament has highlighted the potential bias and discrimination derived from the use of AI applications, especially regarding the algorithms in which these applications are based on. The tension between efficiency and procedural guarantees is evident; while AI can speed up procedures, it exist the risk that such swiftness compromises fundamental rights such as presumption of innocence and the right to a fair trial.

To mitigate these risk, it is essential to establish regulatory frameworks that secure transparency, accountability and human supervision in AI applications in the criminal field. The Office of the United Nations High Commissioner for Human Rights has emphasized that AI technologies must be grounded in human rights, and those who cannot operate following the international normative must me prohibited or suspended until the adequate protection measures are implemented.

3 Challenges for Human Rights

3.1 *Privacy: Massive Vigilance and Data Protection*

Regarding massive vigilance systems and the management of personal data, AI has generated significant worries in relation to the protection of human rights; although they offer benefits in terms of security and efficiency, they raise critical issues that require in-depth analysis and proper regulation.

One of the main challenges lies in the capability of AI to perform massive vigilance without precedents, through advanced facial recognition systems and behavior analysis which allow to monitor and analyze the activity of individuals in real time, which can turn into an erosion of privacy and the restriction of fundamental freedoms. The compilation and processing of large volumes of personal data by the hand of these systems poses significant risks to the data protection, the lack of transparency in algorithms and the opacity in data handling may make it difficult for individuals to understand how their data is being used and for what purposes. This could lead to a violation of the right to privacy and potential abuses due to algorithmic discrimination. Studies by Brayne (2017), Gangi (2023) highlight the implementation of mass surveillance systems as a driver for the normalization of state surveillance, making it necessary to address the challenges posed by these technologies from a human rights perspective, in order to strengthen the protection of human rights in the digital age.

3.2 *Due Process: Transparency and Explanations for Automated Decisions*

The automation of judicial decisions through algorithms can compromise essential principles such as the presumption of innocence, the right to a fair trial and equality before the law. The opacity of these systems hinders the comprehension of the criteria used in decision-making, which also can infringe the right to effective legal representation. The lack of transparency in AI algorithms prevents defendants and their legal representatives from understanding how evidence has been evaluated or guilt has been determined; this undermines the capacity to contest decisions rooted in algorithmic mistakes or biases, violating the principle of contradiction and the right to an effective legal recourse.

The implementation of AI in the judicial system poses additional questions about responsibility and accountability. In the case of judicial errors derived from algorithmic decisions, it would be complex to determine the responsibilities: the developer of the algorithm, the entity who implemented it or the judge that trusted its results. This ambiguity could generate a lack of trust in the judicial system and violate the right of the people to be judged by competent and impartial tribunals. The use of AI in judicial proceedings may affect the equality of arms between the parties; institutions with greater resources can access more advanced technologies, allowing them to gain undue advantages in litigation, which violates the principle of equality before the law and may result in unfair decisions, thereby increasing power and resource disparities within the judicial system.

To mitigate these challenges, it is essential to develop legal and ethical frameworks that regulate the use of AI in the judicial system, the implementation of human rights impact evaluations, ensure transparency in algorithms and maintain human oversight in decision-making. The European Union has proposed regulations that restrict the use of AI systems in judicial proceedings, emphasizing the need to safeguard citizens' fundamental rights.

3.3 Algorithmic Discrimination: Biases in the Application of AI

As previously mentioned, one of the main issues derived from the application of AI is the risk of perpetuating and amplifying existing inequalities. This stems from the fact that AI is trained on historical data, which may internalize and replicate biases present or preexisting in said data.

A study conducted by the New York-based news agency *Pro-Publica* revealed that the COMPAS software, used in the United States to assess the likelihood of recidivism, tended to overestimate the risk for African American individuals and underestimate it for Caucasian individuals. Such biases not only compromise the equity of the judicial system but also violate fundamental rights such as the presumption of innocence and the right to a fair trial. The lack of transparency in automated decision-making processes prevents those affected from understanding the reasons behind certain determinations, which violates the right to information and limits the ability to challenge unfair decisions. It is crucial to recognize that biases in AI are not merely technical flaws, they are reflections of preexisting structural inequalities in society. Therefore, addressing these challenges requires a multidisciplinary approach that combines technical solutions with social and legal reforms to ensure proper governance, cooperation to defend rights, and the reduction of inequalities in the context of AI (Ricaurte 2024).

4 Regulation and Proposed Solutions

4.1 Legal Gaps in Spanish and European Regulations

The rapid evolution of AI has outpaced the ability of Spanish and European regulations for its adequately development and application, creating significant gaps that affect legal certainty and the protection of fundamental rights. One of the main issues lies in the lack of a clear and precise definition of what constitutes AI. In this regard, the European Union's proposed AI regulation offers a broad definition that could lead to diverse interpretations, making it difficult to identify and regulate specific systems. This ambiguity impacts decision-making regarding civil and criminal liability, which is particularly concerning in sectors such as autonomous transportation and industrial automation (Xavier 2023).

If an autonomous vehicle causes an accident, for example, there is no clear regulation determining whether the responsibility lies with the manufacturer, the vehicle owner or the software developer, which can complicate victim compensation and the proper allocation of liability (EL Kady 2025). Additionally, the growing ability of AI to influence human behavior through subliminal persuasion techniques or information manipulation raises serious ethical and legal concerns. Emerging concepts such as neuro-rights, which aim to protect mental integrity and individual autonomy, are not yet included in European legislation, leaving a gap in protection against potential abuses that could arise from advanced technologies like facial recognition and mass surveillance (Alonso 2021).

The issue of data protection and privacy is a gap, an area where current regulations, such as the General Data Protection Regulation (GDPR), are insufficient in the face of the challenges posed by AI. AI systems process large volumes of personal data with the goal of optimizing their functioning, often without the explicit consent of individuals, which could seriously compromise their privacy. As Mandalapu et al. (2023) points out, AI's ability to generate deepfakes further complicates the protection of individuals' image and reputation, especially in judicial and public security contexts (Delgado 2023).

In the field of surveillance, the use of facial recognition systems by public and private entities has sparked another controversy, as these technologies operate in a regulatory loophole that fails to establish clear boundaries between the balance of public security and the protection of individual's rights (Delgado et al. 2019).

The lack of specific regulation in this area could lead to invasive practices that infringe on fundamental freedoms, such as freedom of expression and the right to privacy. The European Union has begun debating regulations to limit the use of these technologies in public spaces, but the lack of consensus around their effective implementation remains an obstacle to the implementation of appropriate regulations.

4.2 Who is Responsible for Crimes Facilitated by AI?

The attribution of responsibility in crimes facilitated by AI represents a legal challenge of increasing complexity. Traditionally, criminal law has focused on attributing criminal conduct to individuals or legal entities with the capacity for action and intent. However, the emergence of AI systems capable of operating autonomously and learning from their environment has created situations where there is no direct human involvement in the commission of the crime, complicating the identification of the responsible party (El Kady 2025).

An illustrative example is the case that occurred in Hungary in 2019, where an AI was used to simulate the voice of a CEO and request an urgent deposit from a company to transfer a significant sum of money (Hildebrandt 2022). This incident raises questions about who should be held responsible: the developer of the AI, the user who implemented it, or the entity itself that fell victim to the deception due to inadequate security measures. According to El Kady (2025), the increasing autonomy of algorithms complicates the application of traditional principles of criminal law, as the unpredictability of AI disrupts the classic concepts of intent and negligence, essential elements for attributing liability. From a legal perspective, criminal liability in these cases could be analyzed from different angles. On one hand, the responsibility of AI developers could be considered if it is proven that they designed the system with flaws that facilitated its criminal use. However, this attribution would require demonstrating gross negligence or intent in the design, which is complex due to the autonomous and evolving nature of AI. On the other hand, users who employ AI for criminal purposes could be considered direct perpetrators of the crime, especially if there is clear evidence of intent to use the technology to commit unlawful acts. An example of this is the use of deepfakes to create non-consensual sexual content, as occurred in Spain in 2023, where minors manipulated images of schoolmates using AI, sparking a debate on the criminal classification of such behaviors (El País 2024a, b).

According to González-Cuellar (2023), the proliferation of these crimes is evidence of the need to reformulate the concept of authorship in criminal law, as AI has altered the relationship between the individual and the criminal act, blurring the traditional boundaries of responsibility. The issue becomes even more complex when AI acts unpredictably, without specific programming to commit crimes. In these cases, the attribution of responsibility becomes unclear, as neither developers nor users can fully foresee or control the actions of AI. Some legal scholars propose the creation of a regulatory framework that establishes objective liability for operators of AI systems, similar to the liability of legal entities, where they would be required to implement appropriate control and supervision measures to prevent criminal behavior (IUS Magazine 2022).

This approach seeks to balance the promotion of technological innovation with the need to protect society from the potential risks associated with AI. In this regard, Wachter et al. (2018, 2021) has advocated for the importance of introducing mandatory audits for AI algorithms in critical sectors to ensure that systems are not used maliciously or for criminal purposes. Another relevant aspect is the consideration of AI as a potential legal subject. Although AI currently lacks legal personality and therefore cannot be held criminally responsible, there has been debate about granting it a special legal status that would allow for some attribution of responsibility. Hildebrandt (2022) notes that this approach faces significant challenges, as it would imply recognizing AI as an entity with the capacity for action and intent, which contradicts its nature as a tool created and controlled by humans (Morán 2021).

Sanctioning an AI would lack practical meaning, as it does not experience suffering nor can it be deterred through penalties. Therefore, responsibility should fall on the individuals or entities that design, implement, and oversee these systems.

The idea of granting legal autonomy to AI is a fallacy, as ultimately, any decision made by an algorithm is conditioned by the rules established by its creator (Luque et al. 2024).

Current regulation in many countries, including Spain, does not specifically address criminal liability for crimes facilitated by AI. The Spanish Penal Code, for example, does not provide for the attribution of responsibility to AI systems, creating a legal void in situations where technology plays a central role in the commission of the crime (El Derecho 2024).

This regulatory loophole highlights the need to update the legal framework to address the challenges posed by AI in the criminal field, establishing clear criteria for the attribution of responsibility and the necessary preventive measures. As Pagallo (2013) points out, the key to addressing this issue lies in designing regulations that not only establish sanctions but also incentivize good practices in AI development, promoting the implementation of control and supervision mechanisms that minimize risks. Determining liability in crimes facilitated by AI requires a multidisciplinary approach that combines legal analysis with a technical understanding of the technology. It is essential to develop a regulatory framework that sets clear obligations for developers, operators and users of AI systems, ensuring that appropriate safety and surveillance measures are implemented (Mazurier et al. 2019).

Likewise, it is crucial to promote research and academic debate on this topic, promoting collaboration among legal scholars, engineers and ethics experts to comprehensively address the challenges posed by AI in the criminal field. As Martino and Merenda (2021) conclude, effective regulation of AI should not only focus on ex post liability assignment but must also prioritize prevention and accountability at all stages of the development and use of these systems.

4.3 Proposal for Algorithm Supervision and Auditing

The supervision and auditing of algorithms in AI systems are essential processes to ensure that these technologies operate ethically, transparently, and in compliance with current regulations. Algorithmic auditing involves a comprehensive evaluation of algorithms, covering everything from the type of data they use to the impact of their decisions on different social groups. This process not only verifies legal compliance but also identifies risks associated with their application, enabling corrections to be implemented when necessary (Contreras 2026).

A concrete example of algorithmic auditing can be found in the realm of hotel search engines, where the Australian Competition and Consumer Commission (ACCC) conducted an audit of a popular hotel search engine and discovered that the algorithm unfairly favored hotels that paid higher commissions in its ranking system. This finding allowed the ACCC to intervene and demand modifications to the algorithm to ensure fair competition and protect consumer rights (Contreras 2026).

In Spain, significant steps have been taken towards AI supervision with the creation of the Spanish Agency for the Supervision of Artificial Intelligence (AESIA).

This autonomous entity aims to supervise, advise, and provide training on the proper use and development of AI systems, with a particular focus on algorithms (Payá and Delgado 2017; Guthrie 2017). In addition to inspection and sanctioning functions, AESIA seeks to minimize the risks associated with the use of these technologies, ensuring their ethical and responsible development.

4.4 Proposal for the Use of AI as a Support Tool and Not as the Sole Judicial Criterion

It is widely accepted that AI should serve as a support tool for judicial systems, complementing their work without replacing human judgment. This perspective is based on the need to maintain impartiality, equity, and justice in judicial processes, values that are inherently human and which AI, on its own, cannot guarantee.

An example illustrating the risks of relying exclusively on AI in judicial decisions can be seen in the study by Gangi (2025), which highlights that a judge in Colombia used ChatGPT, an AI tool, to resolve a legal action related to the rights of a child with autism. Although the final decision was favorable to the plaintiff, the use of AI without proper supervision sparked controversy and led the Constitutional Court to emphasize that judges should not delegate their responsibilities to AI systems and must disclose their use in judicial processes. This incident underscores the importance of using AI as a complement in the judicial process, enhancing efficiency in analyzing large volumes of information and identifying patterns that can be useful for judges.

However, the final decision must always rest within the judge, who possesses the ability to interpret the law with sensitivity and an understanding of the human context, elements that AI cannot yet replicate. Additionally, it is essential to establish clear regulatory frameworks that guide the use of AI in the judicial sphere, ensuring that its implementation respects fundamental rights and maintains the integrity of the justice system (Payá 2023; Rodríguez et al. 2023; Sanz et al. 2024).

5 Conclusions

The emergence of AI in criminal law has represented a turning point in the evolution of contemporary legal systems. Its application has enabled unprecedented advances in crime prediction, the management of digital evidence, and the optimization of judicial processes, profoundly transforming the administration of justice.

The automation of certain procedures and the analytical capacity of AI systems have sped up decision-making and enhanced the ability of legal operators to manage large volumes of information efficiently. Nevertheless, the increasing reliance on these systems poses structural challenges that transcend the technological realm and directly affect the very configuration of criminal law. The principles of culpability, the presumption of innocence, and the proportionality of penalties, fundamental pillars of criminal theory, could be eroded if AI is not implemented within a framework of guarantees and with the necessary institutional controls (Thamer et al. 2024).

The speed and efficiency promised by these systems cannot replace the rigor and prudence inherent in judicial work, as criminal law is not only based on the rational analysis of facts but also on an ethical and humanistic dimension that AI, by its very nature, has so far been unable to replicate. It is imperative to reflect on the urgent need to establish a balance between technological development and the protection of human rights (Grigore 2022).

AI, although endowed with extraordinary predictive and analytical capabilities, must not become a mechanism that undermines the procedural guarantees of individuals or compromises equality before the law. The automation of certain procedures, if not subjected to adequate control, could generate discriminatory or arbitrary decisions, disproportionately affecting certain vulnerable groups and eroding trust in the administration of justice. Criminal justice cannot become a mechanistic system where algorithms determine the fate of individuals without considering the complexity of each case and the uniqueness of individual circumstances.

Technological advancements must be subordinated to an ethical framework that ensures AI is not used as an instrument of exclusion or as a tool that reproduces structural biases present in society. Equity and proportionality must prevail over any criterion of algorithmic efficiency, as criminal law is not solely a matter of calculation and optimization but a normative system that must guarantee the dignity and freedom of all citizens.

The need for ethical and rights-based regulation is not merely a technical imperative but an essential mandate for the preservation of the rule of law. The incorporation of AI in the criminal sphere should not be interpreted as an inevitable phenomenon that must be accepted without reservations, but as a process that demands rigorous normative construction and a profound debate on its implications. The transparency of algorithms, the auditability of systems and the accountability of their developers must be part of the regulatory framework governing their implementation; otherwise, the risk remains of consolidating a model of justice in which judicial decisions are opaque and unappealable, directly undermining the fundamental principles of due process.

Criminal law, as the ultimate expression of the State's punitive power, cannot delegate its function to systems that lack moral sensitivity nor critical capacity, as criminal judgment is not merely a matter of factual determination but an act of deliberation that must weigh values, contexts, and principles.

Therefore, the regulation of AI in criminal law must be based on a rights-based vision, where technology serves as a complement rather than a substitute for human reasoning, and where fundamental rights are the guiding principle of any legal innovation.

References

ALONSO SALGADO, Cristina (2021). Acerca de la inteligencia artificial en el ámbito penal: especial referencia a la actividad de las fuerzas y cuerpos de seguridad. *IUS ET SCIENTIA*. Vol. 7 Nº 1. pp. 25–36 https://doi.org/10.12795/IETSCIENTIA.2021.i01.03

BALCELLS, Marc. (2020). Luces y sombras del uso de la inteligencia artificial en el sistema de Justicia penal. En A. Cerrillo I Martínez y M. Peguera Poch (Eds.). Retos jurídicos de la inteligencia artificial. Cizur Menor (Navarra): Aranzadi.

BRAYNE, Sarah. (2017). Big Data Surveillance: The Case of Policing. American Sociological Review, 82(5), 977–1008. https://doi.org/10.1177/0003122417725865

BUOLAMWINI, Joy., & GEBRU, Timnit. (2018). Gender Shades: Intersectional Accuracy Disparities in Commercial Gender Classification. Science Advances, 4(1), eaas8576. Proceedings of Machine Learning Research. 81:77–91. Available from https://proceedings.mlr.press/v81/buolamwini18a.html

CHESNEY, Robert., & CITRON, Danielle. (2019). Deepfakes and the New Disinformation War: The Coming Age of Post-Truth Geopolitics. Foreign Affairs, 98(1), 147–155. Available at: https://scholarship.law.bu.edu/shorter_works/76

CHÉN, Oliver. (2022). Uniting Machine Intelligence, Brain and Behavioural Sciences to Assist Criminal Justice. arXiv preprint arXiv:2207.01511. https://doi.org/10.48550/arXiv.2207.01511

Contreras, V. (2026). War without accountability. AI and the fragmentation of responsibility in contemporary warfare: power infrastructures and technological sovereignty. Cuadernos de RES PUBLICA en derecho y criminología, 1–32. https://doi.org/10.46661/respublica.13315

DELGADO MORAN, Juan. José, MAZURIER, Pablo. Andrés. & PAYA SANTOS, Claudio. Augusto. (2019). The race to securitize the arctic in a post-cold War scenario. *Revista de Pensamiento Estratégico y Seguridad CISDE*, 4(1), 59–64. http://hdl.handle.net/10272/17180

DELGADO MORÁN, Juan José. Políticas públicas de seguridad en España. Análisis desde perspectivas criminológicas. *Revista Opinião Jurídica, Fortaleza*, v. 21, n. 37, p. 183- 211, maio/ago. 2023. 1806–0420. https://doi.org/10.12662/2447-6641oj.v21i37.p183-211.2023

El Derecho. com (2024). Responsabilidad penal y la inteligencia artificial en España. Editorial Jurídica El Derecho

El País. (2024). El caso de deepfakes en España reabre el debate sobre la regulación de la IA. Recuperado de https://www.elpais.com

El País. (2024). Una demanda de un niño con autismo abre el debate para regular el uso de la IA en la justicia colombiana. Recuperado de https://elpais.com/america-colombia/2024-08-25/una-demanda-de-un-nino-con-autismo-abre-el-debate-para-regular-el-uso-de-la-ia-en-la-justicia-colombiana.html

European Commission. (2021). Proposal for a Regulation of the European Parliament and of the Council Laying Down Harmonised Rules on Artificial Intelligence (Artificial Intelligence Act) and Amending Certain Union Legislative Acts. COM/2021/206 final

EL KADY, Ramy. (2025). "Challenges of Criminal Liability for Artificial Intelligence Systems." In Exploration of AI in Contemporary Legal Systems, edited by Halim Bajraktari, 1–42. Hershey, PA: IGI Global. https://doi.org/10.4018/979-8-3693-7205-0.ch001

Gangi-Guillén, G. K. (2025). Derechos humanos y derecho penal en la era de la inteligencia artificial: retos y propuestas. Cuadernos de RES PUBLICA en derecho y criminología. https://doi.org/10.46661/respublica.11635

GANGI GUILLEN, Giuseppe. Kodjack. VII. (2023). Dinámicas migratorias en la frontera colombo-venezolana y su relación con la criminalidad transnacional. Revista Científica General José María Córdova, 21(44), 907–924. https://doi.org/10.21830/19006586.984

GARVIE, Clare., BEDOYA, Alvaro. & FRANKLE, Jonathan. (2016). The Perpetual Line-Up: Unregulated Police Face Recognition in America. Georgetown Law Center on Privacy & Technology. https://www.perpetuallineup.org/

GRIGORE, Andrea. Elena. (2022). Derechos humanos e inteligencia artificial. *IUS ET SCIENTIA*, 8(1), 164–175. https://doi.org/10.12795/IETSCIENTIA.2022.i01.10

GUTHRIE FERGUSON, Andrew. (2017). The Rise of Big Data Policing: Surveillance, Race, and the Future of Law Enforcement. New York: NYU Press. https://doi.org/10.18574/nyu/9781479854608.001.0001

HERNÁNDEZ GIMÉNEZ, María. (2019). Inteligencia artificial y Derecho penal. *Actualidad Jurídica Iberoamericana,* 10(bis), 792–843. https://revista-aji.com/wp-content/uploads/2019/06/792-843.pdf

HILDEBRANDT, Mirelle. (2022). Law for Computer Scientists and Other Folk. Oxford University Press. https://doi.org/10.1093/oso/9780198860877.001.0001

LIZ RIVAS, Lenny. (2024).Violencia y agresión entre iguales a través de las TICS: Cyberbulling. AlmaMater. Cuadernos de Psicosociobiología de la Violencia: Educación y Prevención, nº 5, 2024, Dykinson, pp. 89–105. https://doi.org/10.14679/3314

LUQUE JUÁREZ, José. María., SANZ GONZÁLEZ, Roger., PAYÁ SANTOS, Claudio. Augusto & RODRÍGUEZ GONZÁLEZ, Víctor. (2024). Aplicaciones de la inteligencia artificial en la criminología y ciencias policiales. https://doi.org/10.13140/RG.2.2.11118.63041

MANDALAPU, Varun., ELLURI, Lavanya., VYAS, Piyush., & ROY, Nirmalya. (2023). Crime Prediction Using Machine Learning and Deep Learning: A Systematic Review and Future. https://doi.org/10.48550/arXiv.2303.1631010.1109/ACCESS.2023.3286344

MARTINO, Luigi. & MERENDA, Federica. (2021). Artificial intelligence: A paradigm shift in international law and politics? Autonomous weapon systems as a case study. in: Giampiero Giacomello & Francesco N. Moro & Marco Valigi (ed.), Technology and International Relations, chapter 5, pages 89–107, Edward Elgar Publishing. https://doi.org/10.4337/9781788976077.00012

MAZURIER, Pablo, Andrés., DELGADO MORÁN, Juan, José & PAYA SANTOS, Claudio, Augusto. (2019). Gobernanza constructivista de la internet. *Teoría y Praxis*, 17(34), 107–130. https://doi.org/10.5377/typ.v1i34.14823

MIRÓ LLINARES, Fernando. (2018). Inteligencia artificial y justicia penal: más allá de los resultados lesivos causados por robots. *Revista de Derecho Penal y Criminología*, (20), 87–130. https://doi.org/10.5944/rdpc.20.2018.26446

MORÁN ESPINOSA, Alejandra. (2021). Responsabilidad penal de la Inteligencia Artificial (IA). ¿La próxima frontera? *Revista del Instituto de Ciencias Jurídicas de Puebla*, México. vol. 15, No. 48. Pp. 289–323. https://doi.org/10.35487/rius.v15i48.2021.706

Oficina del Alto Comisionado de las Naciones Unidas para los Derechos Humanos. (2021). Los riesgos de la inteligencia artificial para la privacidad exigen una acción urgente. Recuperado de https://www.ohchr.org/es/press-releases/2021/09/artificial-intelligence-risks-privacy-demand-urgent-action-bachelet

PAGALLO, Ugo (2013). The Laws of Robots: Crimes, Contracts, and Torts. Dordrecht: Imprint: Springer.

PASQUALE, Frank. (2015). The Black Box Society: The Secret Algorithms That Control Money and Information. Cambridge: Harvard University Press. https://doi.org/10.4159/harvard.9780674736061

PAYÁ SANTOS, Claudio. Augusto. (2023). El desempeño de la inteligencia en España en el ámbito público, empresarial y académico. *Revista Científica General José María Córdova, 21*(44), 1029–1047. https://doi.org/10.21830/19006586.1222

PAYÁ SANTOS, Claudio. Augusto, & DELGADO MORÁN, Juan. José. (2017). Uncertainty of dimensional analysis of intelligence. *URVIO Revista Latinoamericana de Estudios de Seguridad* 21, 225–239.https://doi.org/10.17141/urvio.21.2017.2962

RICAURTE, Paola. (2024). Las grandes compañías tecnológicas son aliadas de gobiernos autoritarios. El País. Recuperado de https://elpais.com/america/lideresas-de-latinoamerica/2024-10-16/paola-ricaurte-las-grandes-companias-tecnologicas-son-aliadas-de-gobiernos-autoritarios.html

RODRÍGUEZ GONZÁLEZ, Víctor., PAYÁ, SANTOS, Claudio, Augusto., & PEÑA HERRERA, Bernardo. (2023). Estudio criminológico del ciberdelincuente y sus víctimas. *Cuadernos de RES PUBLICA en Derecho y criminología,* (1) 95–107. https://doi.org/10.46661/respublica.8072

SANZ GONZÁLEZ, Roger, LUQUE JUÁREZ, José María, MARTINO, Luigi, LIZ RIVAS, Lenny, DELGADO MORÁN, Juan José, & PAYÁ SANTOS, Claudio Augusto. (2024) Artificial Intelligence Applications for Criminology and Police Sciences. *International Journal of Humanities and Social Science.* Vol. 14, No. 2, pp. 139–148. https://doi.org/10.15640/jehd.v14n2a14

SCANTAMBURLO, Teresa, Andrew CHARLESWORTH, & Nello CRISTIANINI, (2019). Machine Decisions and Human Consequences, in Karen Yeung, and Martin Lodge (eds), Algorithmic Regulation. Oxford Academic. arXiv https://doi.org/10.48550/arXiv.1811.0674710.1093/oso/9780198838494.003.0003

THAMER NAJM, Abdullah Abbas., HAMEED, Raed., AKRAM KADHIM, Ali., & HASHIM QASIM, Namer. (2024). Artificial intelligence and criminal liability: exploring the legal implications of aIenabled crimes. *ENCUENTROS. Revista de Ciencias Humanas, Teoría Social Y Pensamiento Crítico.,* 22 140–159. https://doi.org/10.5281/zenodo.13386675

WACHTER, Sandra., MITTELSTADT, Brent., & RUSSELL, Chris. (2018) "Counterfactual Explanations Without Opening the Black Box: Automated Decisions and the GDPR", Harvard Journal of Law and Technology. 31 (2) 841–887. https://doi.org/10.2139/ssrn.3063289

WACHTER, Sandra., MITTELSTADT, Brent., & RUSSELL, Chris. (2021). Why fairness cannot be automated: Bridging the gap between AI and human rights law. Computer Law & Security Review, 43(4), 34–55. https://doi.org/10.1016/j.clsr.2021.105567

XAVIER JANUÁRIO, Tulio. Felippe. (2023). Inteligencia artificial y responsabilidad penal de personas jurídicas: un análisis de sus aspectos materiales y procesales: Un análisis de sus aspectos materiales y procesales. *Estudios Penales Y Criminológicos,* 44(Ext.), 1–39. https://doi.org/10.15304/epc.44.8902

Cybercrime and Artificial Intelligence: Legal and Technological Challenges

Adrián Nicolás Marchal González

1 Introduction

The current landscape of information security holds critical importance in our digital environment. The tools used in both professional and social contexts have evolved substantially, transcending email to encompass messaging platforms, collaboration tools, cloud applications, and file-sharing services (Elbayadi, 2021). This diversification has considerably expanded the attack surface and fragmented the security ecosystem, which frequently comprises isolated solutions, thereby increasing operational costs and the probability of security breaches. Concurrently, cyber threats have undergone significant evolution in terms of sophistication and omnipresence. Cybercriminals actively exploit this expansion of the attack surface, utilizing collaborative tools, supply chains, and human factors to circumvent defensive mechanisms and compromise accounts. In this dynamic and increasingly complex context, artificial intelligence has emerged as a transformative technology with growing impact across multiple sectors. The convergence between AI and cybersecurity currently constitutes a fundamental field of study, given AI's dual potential: both to enhance the offensive arsenal of cybercriminals and to strengthen defensive capabilities. Understanding this duality is essential to effectively navigate the complex landscape of current and future threats.

The increasing complexity of cybercrimes drives the need to implement more robust and intelligent cybersecurity methods. AI, encompassing various technologies such as Machine Learning, Deep Learning, and Natural Language Processing, offers the potential to develop defensive mechanisms capable of making real-time decisions against sophisticated attacks. However, in parallel, malicious actors are inevitably incorporating AI techniques into their arsenal to operate in cyberspace and cause

A. N. Marchal González (✉)
Department of Security and Defense, Nebrija University, Madrid, Spain
e-mail: amarchal@nebrija.es

L. A. Garcia-Segura (ed.), *Collaborative Insights on Technology and Legal Frameworks*, Advanced Sciences and Technologies for Security Applications,
https://doi.org/10.1007/978-3-032-27714-5_10

significant damage without being detected. The use of AI and automation enables them to execute more elaborate social engineering attacks at a larger scale. AI algorithms are used to identify suitable targets, develop convincing profiles, create plausible scenarios to manipulate victims, and adapt strategies dynamically. This provides attackers with a repertoire of AI-driven tactics against which conventional defenses are frequently ineffective. Generative AI, particularly, is being employed to design more sophisticated and personalized attacks, with its growing use in ransomware development being especially concerning, featuring large language models (LLMs) specifically designed to maximize the impact of such attacks.

This technological advancement introduces new challenges in the security domain. The integration of AI into information systems entails inherent risks, as these systems present specific vulnerabilities such as algorithmic biases, susceptibility to adversarial attacks (through evasion and data poisoning techniques), and manipulation of training data. These vulnerabilities can compromise the reliability and integrity of AI-based systems (Howard, 2023). Adversarial attacks against AI models, including data manipulation and poisoning, as well as the exploitation of algorithmic vulnerabilities, substantially expand the spectrum of cybersecurity risks. The reliability of these systems, understood as confidence in their behavior according to established parameters, depends on technical and socio-technical characteristics and principles such as accuracy, robustness, explicability, privacy, security, bias management, impartiality, responsibility, and transparency. The protection of fundamental assets such as data (raw, training, testing), models (algorithms, parameters), artifacts (frameworks), actors (data owners, scientists, engineers), and processes (collection, processing, training) is crucial. Interestingly, although 66% of surveyed professionals consider that AI will have the greatest impact on cybersecurity in the coming year, only 37% have implemented processes to evaluate its security before deployment.

Despite the growing relevance of AI in cybersecurity, there is a lack of comprehensive analyses on artificial intelligence methods employed to combat cybercrimes. Additionally, the risks associated with AI implementation, such as the malicious use of algorithms or the intrinsic vulnerabilities of machine learning systems, have not been exhaustively explored in scientific literature. Existing studies show a notable bias, predominantly focusing on intrusion detection and prevention systems (IDPS), with the Support Vector Machines (SVM) technique prevailing. A concentration of publications in just two main sources has also been identified (Watters, 2024). This situation reveals a knowledge gap regarding the complete landscape of AI-based threats and defenses, suggesting the need to adopt more recent techniques and diversify publication channels to intensify research in AI applied to cybersecurity. Academic research must address this emerging issue at the intersection of artificial intelligence and information security, identifying trends and gaps, and serving as a reference for future studies on the balance between technological innovation and security. Investigating the emerging threat of AI-enhanced cyberattacks is crucial to understand their current and future level of sophistication.

The main objective of this chapter is to analyze the security threats linked to the integration of artificial intelligence in information systems, as well as to examine AI applications to strengthen defensive capabilities. It aims to identify emerging vulnerabilities and risks derived from the offensive use of AI by cybercriminals, analyze the intrinsic threats to AI systems themselves, and present how this technology is applied in cyber defense. The research aspires to provide a comprehensive framework for understanding current challenges and anticipating future threats. This involves examining how AI acts as a catalyst and sophisticator of cyberattacks, analyzing the main AI-based offensive tactics, and considering the malicious use of generative AI. Likewise, the specific vulnerabilities of AI systems, adversarial attacks, data manipulation, and risks associated with algorithmic biases will be addressed. Finally, the defensive applications of AI, the enhancement of traditional defense mechanisms, and the challenges inherent to the implementation of AI-based defensive strategies will be detailed. The chapter will also outline strategies and best practices to mitigate these threats and strengthen cybersecurity in the AI era.

This study has significant relevance both at a theoretical and practical level. On the theoretical plane, it contributes to the development of the research field at the intersection of artificial intelligence and information security, an area that demands comprehensive analyses. By identifying trends and knowledge gaps, it establishes a reference base for future research. It allows for a deeper understanding of how AI is transforming the nature of cyberattacks and defenses, exploring the complex interaction between human and technological factors. In the practical realm, the chapter seeks to provide researchers and professionals with a deeper understanding of current methods to ensure cybersecurity against the growing complexity of cybercrimes (Golden, 2025). It provides crucial information for organizations on emerging threats and the possible limitations of existing defensive infrastructures, indicating the need to invest in AI-based cybersecurity systems and implement good practices. It highlights the importance of a balanced approach between human factors, technological factors, and policies in organizational security strategies, emphasizing the need for a multi-layered defense that includes updated systems, network segmentation, real-time monitoring, backups, and continuous user training. It addresses the need to evolve cybersecurity awareness beyond basic training toward effective behavioral change to counter advanced social engineering attacks.

The structure of this chapter is organized as follows: after this introduction, the next section will explore how Artificial Intelligence is used in service of cybercrime (offensive AI), detailing the acceleration and sophistication of attacks, the main tactics, and the malicious use of generative AI. Subsequently, the security threats intrinsic to AI systems themselves will be examined, including specific vulnerabilities, adversarial attacks, and the impact of algorithmic biases and errors. Then, we will analyze the role of Artificial Intelligence in cyber defense (defensive AI), presenting its applications to counter threats and the associated challenges. Next, strategies and best practices will be proposed to strengthen cybersecurity against AI-driven threats and to secure systems based on this technology. Finally, we will present the main conclusions derived from the analysis. A list of bibliographic references is included at the end of the chapter to facilitate deeper exploration of the addressed concepts.

2 Artificial Intelligence in the Service of Cybercrime (Offensive Ai)

Artificial Intelligence has revolutionized numerous sectors of human activity. However, its inherent capacity to process external data, learn from it, and apply knowledge through adaptive mechanisms presents a clear duality in the cybersecurity domain (Seiersen, 2023). Although AI constitutes a promising tool for reinforcing defensive mechanisms, it is simultaneously being actively adopted by cybercriminals to refine their methodologies, strategies, and general attack models. This malicious application is generating novel challenges in digital security, transforming the threat landscape toward increasingly complex scenarios.

It is fundamentally necessary to comprehend the vulnerabilities and risks associated with the integration of AI in information systems, since the same technology that enhances security can be utilized for malicious purposes. Various investigations have analyzed the emerging threat of AI-powered cyberattacks, providing valuable information regarding their malicious use. Thirty-two offensive AI capabilities have been identified, evidencing its potential to enhance adversarial techniques. AI is consolidating itself as a key tool that, when maliciously utilized, significantly increases the risks of advanced attacks.

2.1 AI as a Tool for Acceleration and Sophistication of Cyberattacks

The evolution of cyber threats has led to increasingly sophisticated and omnipresent attacks. Cybercriminals are inevitably incorporating Artificial Intelligence techniques to circumvent cyberspace defenses and cause greater damage without detection. AI contributes to this sophistication by leveraging its capabilities to launch more intelligent and adaptive attacks.

One of the key contributions of AI to cybercrime is the acceleration of attacks. Automation, driven by AI systems, allows attackers to execute complex tasks at greater scale and velocity than would be manually possible. This notably expands the threat landscape, as even actors with limited technical expertise can conduct sophisticated operations. Automation also facilitates attack scaling, increasing their probability of success. Legitimate marketing and social media management tools, such as Autopilot and Hootsuite, can be repurposed by cybercriminals to amplify their attacks and systematically increase their impact.

AI also plays a decisive role in the sophistication of cyberattacks. Algorithms can be utilized to identify suitable targets, develop convincing profiles, and create plausible scenarios to manipulate unsuspecting targets (Moschovitis, 2021). This is particularly relevant in social engineering attacks, where criminals exploit human psychology. Since social engineering is fundamentally based on manipulating emotions, instincts, and vulnerabilities, AI's capacity to personalize attacks and generate more convincing messages makes them more difficult to detect and neutralize through conventional security measures. Attackers employ generative AI tools to design more sophisticated and personalized attacks.

Furthermore, AI allows attackers to adapt their strategies in real-time. This provides them with an arsenal of dynamic tactics against which traditional defenses struggle to remain updated. Cybercriminals can adapt more rapidly than implemented security controls.

The "weaponization" of AI for criminal purposes includes its use to generate deepfakes, enhance surveillance, and exploit automated decision-making. The social engineering risk is intensified, leveraging human trust as a highly effective attack vector. AI amplifies the risks of advanced attacks, such as intelligent phishing, adaptive malware, and impersonation techniques via deepfakes (Kshetri, 2021).

In summary, AI enhances cybercriminals' capabilities through:

- Acceleration of the attack process through automation and scalability
- Sophistication of tactics through personalization, creation of convincing content, and real-time adaptation
- More effective exploitation of human vulnerabilities (social engineering)
- General enhancement of adversarial capabilities

Findings from various investigations suggest that current defensive infrastructures may become inadequate to address the increasing velocity and complex decisional logic of AI-powered attacks.

2.2 *Principal Offensive Tactics and Techniques Based on AI*

The integration of Artificial Intelligence into information systems introduces a series of critical threats. These encompass data manipulation and adversarial attacks, malicious utilization of advanced language models, and automated generation of cyberattacks. Threats derived from AI integration include data manipulation, exploitation of algorithmic vulnerabilities, and generation of advanced attacks that considerably expand the spectrum of cybersecurity risks (Coates, 2022).

Below, we detail some of the principal offensive tactics and techniques identified:

2.2.1 Adversarial Attacks

Adversarial attacks represent a significant vulnerability in AI systems. In these attacks, cybercriminals perform minimal alterations to input data to deceive models and provoke erroneous decisions. These alterations, frequently imperceptible to the human eye, are sufficient to confuse the algorithm. These attacks exploit the dependence of AI models on large volumes of data. The alteration of training data can distort the model's learning, compromising its performance and generating risks for system security. Evasion, where attackers compromise algorithm inputs to exploit outputs, constitutes a type of adversarial attack that generates input perturbations denominated "adversarial examples." The vulnerability of AI systems to these attacks represents one of the principal concerns, as they can compromise the integrity and reliability of confidential data protection.

2.2.2 Data Poisoning

Data poisoning is an offensive tactic where hostile actors alter data or the AI model with the objective of modifying algorithmic behavior (Skulmoski, 2022). The purpose is to fulfill criminal objectives such as sabotaging the system or inserting a backdoor. The manipulation of input data constitutes one of the most concerning threats, as it can compromise model integrity by introducing malicious information into the system. Poisoning can occur at any phase of the data lifecycle. These attacks have been specifically documented in the context of image processing models in automotive applications, such as traffic sign detection.

2.2.3 Automated and Sophisticated Generation of Social Engineering and Phishing Attacks

AI and automation allow the launch of more sophisticated social engineering attacks at greater scale. Generative AI tools are utilized by attackers to create more elaborate and personalized attacks. Large Language Models (LLMs) can generate highly convincing texts, adopt specific linguistic styles, and evade traditional security filters. This aggravates the social engineering risk by exploiting human trust. AI can automate the collection of sensitive information from victims. The automated generation of undetectable phishing URLs using techniques such as LSTM (Long Short-Term Memory) has been documented.

2.2.4 Creation of Convincing False Content (Deepfakes and Vishing)

AI-powered tools allow cybercriminals to create hyperrealistic audio and video clips (deepfakes) or simulate realistic telephone calls (vishing) apparently originating from trusted entities. Specific tools such as DeepFaceLab and Descript's Overdub are utilized in this context. These capabilities are applied in telephone phishing attacks or video impersonation, making the deception much more convincing (Rains, 2023). The use of deepfakes is identified as an advanced impersonation technique enabled by AI and as part of the criminogenic potential of this technology. Content generated or modified through AI (images, audio, or videos, denominated "ultrafakes") presents substantial risks.

2.2.5 Intelligent and Adaptive Malware

AI contributes to the development of adaptive malware that can learn from its environment and evade detection. The generation of adversarial malware, which seeks to create undetectable malicious code to circumvent Machine Learning-based detection systems, constitutes an offensive capability that utilizes techniques such as GAN (Generative Adversarial Network). Intelligent malware can infiltrate environmental control systems, initiating indirect cyber "weapons" that simulate involuntary failures in IT infrastructures. Various studies report the use of diffuse models and adaptive malware as a specific vulnerability.

2.2.6 Exploitation of Vulnerabilities in AI-Based Systems and Distributed Environments

The integration of AI in information systems introduces vulnerabilities exploitable by malicious actors. This includes exploitation in distributed environments such as cloud and IoT systems. AI systems operating in these environments are especially susceptible to data interception during transfer, exposing confidential information. Attackers can employ modern techniques to compromise cloud systems. Decentralization, for example in edge AI integration, introduces additional threats such as malicious data attacks that manipulate learned patterns. Risks are also identified in exploiting implementation gaps to manipulate data or compromise privacy. Automation facilitated by AI, although beneficial for defenders, creates new opportunities for attackers to manipulate automated systems through malicious code insertions or deactivation of key defenses. The manipulation of qubits in quantum systems represents a critical vulnerability point potentially exploitable to compromise calculations, given the role of AI algorithms in quantum error correction (Deutsch Franco, 2022).

AI not only enhances individual techniques but can also be utilized for the automated generation of cyberattacks and orchestration of sophisticated operations in multiple stages. Various studies have analyzed the phases of an AI-powered cyberattack, highlighting that a significant portion of identified techniques concentrates on the access and penetration phase, while the remainder is distributed across exploitation, command and control, reconnaissance, and delivery phases. This evidences how AI leverages its capacity to automate processes and evade traditional defenses throughout the entire attack chain. The possibility of automating complex tasks expands the threat landscape, allowing actors without advanced experience to execute sophisticated attacks.

3 Intrinsic Security Threats to Ai Systems

The incorporation of artificial intelligence into information systems infrastructure constitutes a significant amplifying factor of cybernetic risks that warrants detailed analysis (Cuerda Arnau, 2013). Whereas artificial intelligence systems provide advanced capabilities for automation, decision processing, and enhancement of cybersecurity mechanisms, this same technology simultaneously presents a potentially harmful dimension that exponentially increases the probability of materialization of attacks characterized by elevated levels of sophistication and adaptability. The technological complexity inherent to AI systems, coupled with their automation capabilities, substantially expands the spectrum of exploitable vulnerabilities. In the subsequent sections, this chapter examines the specific vulnerabilities consubstantial to artificial intelligence systems and analyzes adversarial techniques, including manipulation and data poisoning, which represent critical threats derived from their implementation.

3.1 Specific Vulnerabilities of AI and its Underlying System

Artificial intelligence systems present specific susceptibilities to security compromises; indeed, they may be subject to manipulation or attacks by malicious agents, thereby compromising their operational functionality and generating new vulnerability vectors. AI is subject to intrinsic vulnerabilities that may undermine the reliability and integrity of the systems where it is implemented (Strang, 2023).

A fundamental vulnerability resides in algorithmic bias. Generative AI systems depend intrinsically on datasets and logical structures provided during their training phase. Any bias or irregularity present in such data may adversely influence the decision-making processes and outcomes of the system, thereby increasing the associated technical risks and ethical considerations. Algorithmic bias can significantly compromise systemic reliability and integrity. The absence of adequate human supervision in these systems may amplify such anomalies. The minimization of biases constitutes a cardinal principle in the ethical framework of AI, while the reduction of algorithmic bias represents a recommended security control for preventing operational dysfunctionalities. The effective management of bias constitutes, likewise, a specific aspect requiring evaluation in the cybersecurity risk analysis of AI systems.

The lack of explicability or interpretability in certain AI models configures another significant vulnerability. This opacity in automated decision-making processes may hinder the identification and rectification of errors or systemic vulnerabilities. The loss of transparency and interpretability has been categorized as a specific threat in the proposed normative framework for AI regulation in the European Union. Explicability and interpretability constitute socio-technical characteristics that fundamentally contribute to the reliability of systems based on this technology.

The structural dependence of AI systems with respect to massive volumes of data represents a vulnerability of significant character. These systems depend integrally on the data and training logic implemented. Deficient management of access mechanisms to critical information or inadequate control of data processed by the model may constitute exploitable vulnerabilities. Additionally, the alteration of training data may significantly distort the learning patterns of the model.

AI systems present, likewise, vulnerabilities derived from the underlying components utilized in their development (Carneiro Pacheco, 2024). The compromise of constitutive elements of an AI application may materialize if hostile actors manage to exploit vulnerabilities in implemented open-source libraries. The utilization of vulnerable components constitutes, per se, a vulnerability. To mitigate the potential level of compromise, it is imperative to implement security policies adequately integrated into the organizational security strategy and technological asset management mechanisms.

A distinctive characteristic of AI systems, particularly those endowed with self-learning capabilities, resides in their temporal evolution. While this property may optimize performance, it simultaneously implies that security properties may experience progressive degradation. This phenomenon underscores the necessity to implement dynamic security tests and continual risk assessment.

The integration of AI in distributed environments such as cloud infrastructures, IoT systems, and AI-based networks introduces specific vulnerabilities. AI systems operative in such environments, such as machine learning in Drone-Cloud configurations, present particular susceptibility to data interception during transfer processes. Cloud services, characterized by their scalability and multi-user model, constitute preferred targets for malicious actors who exploit systemic vulnerabilities. IoT devices require robust access control mechanisms, secured communication protocols, and periodic updates to mitigate exploitation risks. The integration of edge AI in distributed systems presents specific risks of attacks via malicious data that may manipulate learning patterns. The multi-user nature and dependence on massive data volumes render these environments priority targets for malicious actors.

The integration of AI in quantum systems represents an emerging risk of particular relevance. The manipulation and management of qubits constitute a critical vulnerability point. Considering that AI algorithms perform a crucial function in error correction and qubit stabilization, any exploitation of these vulnerabilities by malintentioned agents could compromise the integrity of quantum calculations.

Although AI may enhance the capabilities of network administrators to identify vulnerabilities, the inherent automation of these systems simultaneously generates new opportunities for malicious actors to manipulate automated systems. The malicious insertion of code or neutralization of critical defensive mechanisms constitute significant risks in this context.

Finally, the absence of a security evaluation integrated into the global organizational cybersecurity strategy represents a structural vulnerability. It is noteworthy that, while 66% of subjects surveyed in a reference study consider that AI will exert a significant impact on cybersecurity, only 37% have implemented evaluative security processes prior to the deployment of these technologies.

In connection with these vulnerabilities, significant ethical considerations emerge, including concerns relating to data privacy, excessive control mechanisms, and the potential adoption of erroneous decisions without human intervention. The privacy of data processed by AI systems may be compromised, as may the potential unauthorized disclosure of sensitive personal information.

3.2 Adversarial Attacks, Manipulation and Data Poisoning

Malicious agents are implementing AI-based tools to develop attack vectors characterized by superior levels of sophistication and personalization. The inherent capacity of AI to analyze massive volumes of data and dynamically adapt to emerging threats may be instrumentalized for harmful purposes, significantly increasing associated risk levels. This tendency materializes in specific attack techniques such as adversarial attacks, structured manipulation, and systematic data poisoning.

Adversarial attacks constitute one of the most significant threats identified in specialized literature. They fundamentally consist of the implementation of minimal alterations in input data with the objective of inducing error in AI models, thereby generating incorrect algorithmic decisions. These attacks exploit the structural dependence of models with respect to massive data volumes; the alteration of training sets may significantly distort the learning patterns of the model, compromising its performance and generating significant operational risks. Adversarial attacks may undermine the integrity and reliability of protection mechanisms for confidential data. Academic research has analyzed these attacks in diverse contexts, including the security of convolutional neural networks (such as capsule networks), adversarial environments in reinforcement learning, evasion attack simulation, black-box attacks against smart electrical infrastructures, and evasion models of intrusion detection systems. Mitigation strategies include the implementation of specialized tools for the detection of adversarial examples, adversarial training to increase model robustness, and the utilization of architectures that are not easily transferable.

Data manipulation represents another critical threat in this context. Hongsong et al. (2024) identify it as particularly concerning, noting that manipulation of input data may compromise the structural integrity of AI models. In distributed edge AI configurations, attacks via malicious data that manipulate the patterns learned by autoencoding models have been documented. Inadequate selection or deficient implementation of secured communication protocols, especially in IoT devices, increases exposure to data manipulation vectors, a problematic extensively discussed in the context of AI-based systems. Vulnerabilities present in cloud services may be exploited for data manipulation operations. In advanced natural language processing models, data manipulation constitutes a frequently documented threat.

Data poisoning represents a sophisticated attack technique whereby hostile actors systematically alter data or the AI model. The fundamental objective consists of modifying algorithmic behavior to satisfy harmful purposes, such as system sabotage or the implementation of unauthorized access (backdoors). The alteration of training data significantly distorts the learning patterns of the model. Data poisoning constitutes a particular challenge for anomaly detection systems based on AI, potentially affecting their operational susceptibility. To prevent poisoning attacks, it is imperative to implement component securitization mechanisms throughout the system life cycle (Ciampa, 2024), continually evaluate the cybernetic exposure of the model, strategically expand the volume of the dataset to reduce the potential influence of malicious samples, and implement robust pre-processing mechanisms to purify training data. Data poisoning attacks have been documented in critical applications such as traffic signal detection in autonomous vehicles. The continued analysis of data security throughout its complete life cycle is essential, considering that poisoning operations may be executed at any phase of processing.

In addition to these techniques, AI is enhancing other attack modalities. AI/ML systems can generate personalized attacks that emulate human communication and elude traditional security filtering mechanisms. This capacity significantly exacerbates the risk associated with social engineering techniques, exploiting trust elements in human interactions. Malicious agents implement generative AI tools to develop attack vectors characterized by superior levels of sophistication and personalization, dynamically adapting their strategies in real time. This phenomenon includes advanced modalities such as intelligent phishing, adaptive malware, and sophisticated impersonation techniques via deepfakes. AI facilitates the orchestration of complex multiphase attacks, process automation, and the evasion of conventional defensive mechanisms. The increasing utilization of AI technologies in ransomware development constitutes a concerning trend, anticipating that the implementation of large language models (LLMs) will significantly increase their harmful impact. Denial of service (DDoS) attacks may, likewise, interrupt the normal operability of AI-based systems via unauthorized access. The inherent automation of AI expands the threat spectrum, allowing even actors with limited technical experience to execute operations characterized by elevated levels of sophistication.

The evolution of social engineering attacks, enhanced by the implementation of AI and automation mechanisms, allows malicious agents to personalize communications with superior levels of verisimilitude. These attacks have transcended traditional channels such as email, exploiting new digital communication pathways. Precisely elaborated messages emulate legitimate communications, significantly hindering the distinction between authentic and malicious content. Traditional defensive mechanisms, based exclusively on technical controls, are insufficient against these dynamic tactics enhanced by AI systems.

4 Artificial Intelligence in Cyber Defense (Defensive Ai)

The increasing complexity of cybercrimes has generated a demand for more robust and intelligent cybersecurity mechanisms. Contemporary organizations manage information volumes that require substantial automation. In this context, artificial intelligence emerges as a fundamental technology for strengthening information security, significantly transforming organizational defensive paradigms. AI contributes decisively to process automation, decision optimization, and global enhancement of cybersecurity systems, revolutionizing threat detection and response mechanisms.

Nevertheless, AI presents a functional duality in the cybersecurity ecosystem. While constituting a powerful tool for implementing defensive strategies, it can simultaneously be instrumentalized for malicious purposes by hostile agents to execute attacks characterized by superior levels of sophistication and adaptability.

Consequently, it becomes imperative to develop a profound understanding of AI's defensive applications, its capabilities to enhance traditional mechanisms, and the intrinsic challenges associated with its implementation.

This section systematically examines artificial intelligence applications in the defensive cybersecurity domain.

4.1 AI Applications to Counter Cyber Threats

Artificial intelligence has demonstrated its instrumental value across diverse domains of defensive cybersecurity. It facilitates the automation of identification and response processes against threats, enables the analysis of massive data volumes to predict potentially harmful behaviors, and presents adaptive capabilities against emerging threats.

One area where AI has made significant contributions is in Intrusion Detection Systems (IDS). Numerous investigations have been developed in this field, particularly in their application to IoT networks. Various deep learning and machine learning techniques have been proposed to optimize attack vector detection. Among these, noteworthy approaches include: DNN models with discriminated feature selection, multi-agent architectures implementing deep Q-networks and attention mechanisms, ensemble models integrating bidirectional LSTM networks and DCGAN, and models based on RL algorithms operating in adversarial environments. Other studies have explored implementations based on DBN, DNN, and MLPNN to detect malicious communications or classify traffic in IoT networks, utilizing standardized datasets such as UGR16, UNSW-NB15, and KDD99. The integration of AI into these systems has allowed for significant reduction in computational complexity, optimization of training times, and minimization of false positive rates (Thomson, 2024).

Beyond IDS, AI finds application in other defensive domains:

- **Threat Detection and Behavioral Analysis**: AI facilitates the analysis and replication of human behavioral patterns to perfect threat detection. It enables advanced processing of logs and audits, correlation of security events, and monitoring based on behavior and social network analysis for security intelligence. Anomaly detection through unsupervised learning techniques constitutes a fundamental strategy for identifying emerging threats.
- **Vulnerability Management**: AI optimizes vulnerability assessment and management processes, including automated testing, fuzzing techniques, and penetration evaluations, effectively identifying weak points in technological infrastructures.
- **Malware Detection**: The implementation of AI/ML techniques in malware detection systems represents an increasingly widespread approach.
- **Threat Intelligence**: AI enhances the analysis of sources such as the dark web, covert communications, sentiment analysis, and thematic modeling to obtain actionable intelligence on potential threats.
- **Risk Management**: AI's analytical capabilities are applied in risk assessment, quantification, scoring, and prioritization processes.
- **Authentication and Access Control**: Areas such as multi-factor, contextual, or attribute-based authentication, and role-based access control (RBAC) or attribute-based access control (ABAC) mechanisms significantly benefit from the integration of AI technologies.
- **Incident Response**: AI optimizes detection, identification, classification, categorization, and analysis processes for incidents, facilitating the automation of responses and recovery plans.
- **Forensic Analysis**: Advanced analytical capabilities of AI are particularly valuable in forensic investigations related to security incidents.
- **Asset and Configuration Management**: AI provides significant capabilities in asset management, inventories, configuration, and validation of security controls, allowing for the automation of configuration processes.
- **Data Loss Prevention**: AI-based systems enhance the detection and prevention of sensitive information leaks or losses.
- **Network Security**: Strategies such as microsegmentation can be effectively integrated with AI-based solutions.
- **Recommendation Systems**: AI facilitates the implementation of contextualized security action recommendation mechanisms.
- **Security in Critical Sectors**: AI finds application in areas such as attack detection in smart electrical infrastructures and in automotive systems, specifically in road signaling detection systems. Research has been conducted on the use of AI for predicting transient stability in electrical networks with security optimization objectives. In the healthcare sector, it is fundamental to protect models and data of AI-based systems due to the increase in targeted attacks.

The performance evaluation of AI models in cybersecurity frequently implements standardized metrics such as accuracy (ACC), recall (REC), precision (PRE), and F-measure. These metrics allow for the objective quantification of the effectiveness of AI applications in threat detection and mitigation processes.

4.2 Enhancement of Traditional Defense Mechanisms through AI

The integration of artificial intelligence not only introduces innovative defensive capabilities but also significantly enhances pre-existing security mechanisms. AI is progressively being incorporated into established IT infrastructures, allowing for process automation and optimization of operational efficiency.

The intrinsic capacity of AI to analyze massive data volumes and predict behavioral patterns substantially improves conventional security tools. For instance, Intrusion Detection and Prevention Systems (IDPS), traditional elements in defensive architectures, notably increase their effectiveness when empowered through AI technologies (Vaidya, 2024). The integration of these capabilities perfects threat detection processes by analyzing complex patterns and anomalous behaviors that would remain imperceptible to systems exclusively based on predefined rules or static signatures. Real-time monitoring, a nuclear functionality in multiple security systems (such as SIEM), acquires superior predictive and analytical capabilities through AI integration.

AI-driven automation significantly optimizes operational efficiency in traditional security functions. Tasks such as initial threat detection and response, log analysis, security assessments (vulnerabilities, penetration), and configuration management can be partially or totally automated through AI implementations. This optimization liberates specialized human resources to concentrate on higher complexity activities requiring expert judgment. AI's adaptability against emerging threats and its decision-making capacity in real-time confer to defensive infrastructures a superior dynamism compared to static traditional controls.

Data protection, a fundamental pillar in conventional security architectures, is strengthened through the integration of AI with its predictive and response capabilities against cyberattacks, particularly in systems protecting information stored in cloud environments. This integration decisively contributes to guaranteeing informational integrity and confidentiality.

In synthesis, AI does not replace the need for traditional defensive mechanisms such as firewalls, antivirus, or cryptographic systems, but acts as an intelligent layer that enhances their detection, response, and adaptation capabilities against an evolving threat landscape (Lawrence, 2024).

4.3 Challenges in the Application of Defensive AI and the Need for Human Validation

Despite its transformative potential, the implementation of AI in cyber defense faces multiple significant challenges. The same technology used to strengthen defensive infrastructures can be instrumentalized by hostile agents, configuring the concept of adversarial AI. Attackers systematically implement AI-based tools, including generative technologies, to develop attack vectors characterized by superior levels of sophistication, personalization, and scalability, adapting more rapidly than conventional security controls. Paradigmatic examples include intelligent phishing, adaptive malware, advanced impersonation techniques through deepfakes, and the development of large language models (LLMs) specifically oriented toward criminal activities. AI-powered offensive techniques predominantly concentrate on critical phases such as access and penetration. Existing defensive capabilities will progressively become inadequate against the increasing speed and complexity of AI-driven attacks.

In addition to the need to counter AI-enhanced attacks, defensive systems based on this technology present intrinsic vulnerabilities. AI is subject to specific vulnerabilities that can compromise its reliability and integrity:

- **Adversarial Attacks**: Malicious agents can manipulate the input parameters of AI algorithms to exploit anomalous behaviors in their results (evasion) or systematically alter data or models to modify their behavior (data poisoning). These attacks exploit the structural dependency of models on massive data volumes, distorting their learning patterns.
- **Algorithmic Biases**: The presence of biases or irregularities in training sets or logical structures can adversely influence decision processes and system results, increasing technological risks and ethical considerations. The effective management of these biases constitutes a fundamental requirement to guarantee the reliability of AI-based systems.
- **Model or Data Exposure**: Hostile actors actively seek to leak information about algorithmic architectures, parameters, or training sets.
- **Component Compromise**: The exploitation of vulnerabilities in libraries or open-source modules implemented in the development of AI systems can compromise the global integrity of the system.
- **Operational Dysfunction**: Failures in AI-based applications can materialize through denial of service (DoS) attacks, manipulation of input parameters, or undetected errors in code implementations.

AI systems can evolve temporally through self-learning mechanisms, implying potential degradation of security properties and consequent need to implement continued dynamic evaluations.

A particularly significant technical challenge lies in the security evaluation of AI-based systems. It becomes imperative to implement dynamic risk assessment methodologies, considering that numerous organizations lack structured processes to systematically evaluate the security of AI implementations prior to their deployment. The reliability of AI algorithms fundamentally depends on dynamic evaluations of sources and datasets utilized.

Facing these complex challenges, the critical need for human validation and supervision becomes evident. Exclusively technical controls are insufficient to effectively protect individuals and organizations against attacks characterized by increasing levels of complexity and adaptability (Masys, 2024). The absence of adequate human supervision in AI systems can significantly amplify existing problems, hindering the detection and rectification of potentially harmful errors for users' fundamental rights. Although tools such as anomaly detection through unsupervised learning techniques offer advanced analytical capabilities, they present intrinsic challenges such as false positive generation and susceptibility to data poisoning attacks, requiring complementarity with expert human supervision and robust validation mechanisms.

Specialized training in cybersecurity becomes crucial for organizational personnel, a priority target in numerous attack vectors, to effectively identify social engineering indicators. However, isolated training is insufficient; it must evolve toward paradigms that drive effective behavioral transformations. It becomes imperative to implement integrated approaches that combine technological and formative elements to provide comprehensive protection. This integration implies developing risk management strategies oriented toward users that strengthen human resilience against complex threats.

The increasing sophistication of AI-enhanced cyber threats equally poses significant challenges for security leadership. Information security officers must effectively translate technical risks into quantifiable organizational impacts (financial, regulatory, reputational) to communicate them effectively to governing bodies.

In conclusion, although AI provides defensive capabilities without historical precedent, its implementation must be based on a profound understanding of its inherent vulnerabilities and offensive strategies based on this technology (Jensen, 2022). It becomes indispensable to adopt balanced approaches that integrate advanced AI technologies, optimized traditional defensive mechanisms, significant investment in human awareness and resilience, and robust regulatory and ethical frameworks. Interdisciplinary collaboration and continued dynamic risk assessment constitute essential elements to effectively navigate the complex cybersecurity landscape in the era of artificial intelligence.

5 Strategies and Best Practices to Strengthen cybersecurity

The integration of artificial intelligence into information systems has introduced a set of critical threats that require specific mitigation strategies. Among the main vulnerabilities identified are the systematic manipulation of data, adversarial attacks, malicious instrumentalization of advanced language models, automated generation of offensive vectors, and exploitation of algorithmic and componential vulnerabilities.

The European Union Agency for Cybersecurity (ENISA) proposes a normative framework of specific security controls designed to prevent and mitigate principal threats against AI systems, with particular emphasis on machine learning architectures and generative artificial intelligence. The following constitute recommended practices according to the framework developed by ENISA:

- **Mitigation of evasion attacks**: Malicious agents seek to compromise the input parameters of algorithms to manipulate results, generating "adversarial examples." Mitigating strategies include the implementation of hostile input detection mechanisms, utilization of adversarial training techniques to increase model robustness, and design of architectures that are not easily transferable. Adversarial training represents an emerging approach with empirical validation in research to strengthen the robustness of machine learning models.

- **Prevention of data poisoning**: Attackers systematically alter data sets or algorithmic architectures to modify their operational behavior. Prevention requires comprehensive securitization of components throughout the complete life cycle, continued evaluation of the cyber exposure of the model, strategic expansion of the data set to dilute the influence of malicious samples, and implementation of pre-processing filters to purify training data. These constitute established practices in data hygiene procedures and software life cycle management, adapted to the specific context of AI.
- **Protection against model or data disclosure**: The leakage of sensitive information about algorithmic architectures, parameters, or training sets represents a significant threat. Access control mechanisms must be implemented with maximum technical rigor. The implementation of robust access controls constitutes a consolidated practice in the general field of cybersecurity.
- **Mitigation of component compromise**: The exploitation of vulnerabilities in open-source libraries or other constitutive elements requires application of security policies integrated into the global organizational strategy and technological asset management procedures. The systematic management of vulnerabilities and assets represents a consolidated practice in the field.
- **Prevention of operational failures or deficient functioning**: Dysfunctionalities can originate through denial of service attacks, manipulation of input parameters, or undetected errors in code implementations. Prevention requires minimization of algorithmic biases, continued evaluation to guarantee operational resistance, and design of explainable systems that facilitate identification of vulnerabilities. Algorithmic explicability and bias management represent emerging approaches in development within the ethical and security framework in AI.

Complementary to these specific mitigations, various general strategies are particularly relevant:

- The development of post-quantum cryptography as a preventive measure against future threats derived from quantum computational capabilities that could compromise current cryptographic standards. This emerging approach is based on the anticipation of future computational capabilities.
- The implementation of blockchain technology to mitigate risks in high-impact AI systems, optimizing process traceability, generating immutable records, and guaranteeing informational integrity. Blockchain technology represents an emerging paradigm with significant potential applicability in data and process security.
- The utilization of anomaly detection systems based on AI for identification of cyber threats. Although unsupervised learning constitutes a powerful analytical technique, it presents inherent limitations such as generation of false positives and vulnerability to data poisoning attacks. Operational effectiveness requires

complementation with expert human supervision and robust validation mechanisms. Anomaly detection represents a consolidated and empirically validated practice in cybersecurity.

- In IoT ecosystems, it is imperative to implement robust access control architectures, secured communication protocols, and systematic update and patching procedures. The intrinsic vulnerability of communication protocols in IoT environments constitutes a widely documented risk vector, making their securitization critical. These practices are consolidated in the field of interconnected device security.
- The protection of confidential information in AI-based systems, particularly in distributed environments such as cloud infrastructures, through implementation of advanced cryptography, anomaly detection systems, and real-time monitoring. Data manipulation and exploitation of systemic vulnerabilities represent critical risks in these environments.

A determining factor in the effective mitigation of threats is the human element. Organizational personnel continue to constitute a priority target in attack vectors. Security awareness plays a crucial function. However, to achieve operational efficacy, training programs must transcend basic instruction to generate effective behavioral transformations. Awareness programs must provide personalized guidance based on risk analysis and adaptive training. Isolated training is insufficient to convert personnel into proactive defensive agents, especially considering that empirical studies evidence that a significant percentage of risk behaviors are executed with full knowledge. While awareness and behavioral modification programs constitute fundamental elements, their complementation with protective technological layers is imperative. Organizations cannot base their defensive posture exclusively on technical controls against malicious actors that implement AI and automation to adapt more rapidly than defensive infrastructures. Strengthening human resilience against advanced social engineering techniques is critical. Generative AI facilitates the execution of social engineering attacks characterized by superior levels of sophistication and scalability, through generation of verisimilar scenarios designed to manipulate specific targets.

In the regulatory framework, the European Union AI Act establishes a classification of AI systems according to their risk level (unacceptable, high, minimal) and determines differentiated compliance requirements for providers and users. Systems categorized as high risk (implemented in critical infrastructures, education, employment, healthcare, etc.) must undergo evaluations prior to their commercialization and throughout their operational cycle. Generative AI technologies must satisfy specific transparency requirements, including the explicit revelation of algorithmically generated content, prevention of generation of illegal material, and publication of compendiums relating to protected data utilized during training processes. General-purpose AI models with elevated potential impact and systemic risk must undergo exhaustive evaluations. The regulation expressly prohibits systems classified as unacceptable risk, including cognitive manipulation or social scoring technologies. The prohibition of unacceptable risk systems came into effect in February 2025.

The provisions relating to general-purpose AI systems will be applicable 12 months after the entry into force of the regulation, while high-risk systems will have a period of 36 months to achieve conformity. The regulatory ecosystem continues to evolve.

To adequately document regulatory compliance, organizations must align their security strategies, risk management procedures, and testing methodologies with requirements established in frameworks such as ENISA, NIST, and the EU AI Act. This implies guaranteeing that risk assessments specifically address threats inherent to AI and applicable regulatory requirements, that implemented security controls effectively mitigate said risks, and that security tests validate the efficacy of defenses and global systemic reliability. Interdisciplinary collaboration between cybersecurity specialists, data scientists, and legal/ethical professionals is essential to navigate and document conformity in this dynamic regulatory environment.

In conclusion, the effective mitigation of threats linked to AI requires an integrative approach that combines consolidated practices (access controls, patching procedures, vulnerability management, anomaly detection, securitization of communication protocols), contextualized adaptations of established practices to the specific environment of AI (dynamic risk management, specialized testing methodologies, data purification procedures), and emerging approaches (adversarial training, post-quantum cryptography, blockchain technology, algorithmic explicability). It is fundamental to complement technical measures with a robust approach to the human factor through advanced awareness and behavioral transformation programs.

6 Conclusions

The convergence between Artificial Intelligence and cybersecurity constitutes, in the current socio-technological context, a critical field of study and an existential challenge for contemporary digital society. The systematic analysis of documentary sources reveals a dynamic of extraordinary complexity characterized by the inherent duality of AI, a technology that simultaneously operates as an enhancing instrument in the methodological arsenal of malicious actors (offensive AI) and as a fundamental mechanism for strengthening defensive infrastructures in cyberspace (defensive AI).

The present study has empirically documented how the integration of artificial intelligence and automated processes has structurally transformed the cyber threat ecosystem, providing hostile agents with unprecedented capabilities to exponentially increase both the speed and level of sophistication of their operational methodologies. The inherent capacity of AI systems to process massive volumes of data, identify optimal targets through multivariable criteria, develop highly credible profiles, generate contextually plausible scenarios, and dynamically adapt strategies in real time confers upon malicious actors a repertoire of tactics characterized by a level of dynamism that significantly transcends the capabilities of conventional defensive mechanisms. The emergence of generative AI, in particular, has notably intensified the risk profile in the digital ecosystem, facilitating the automated generation of attack vectors based on social engineering and highly personalized and convincing phishing techniques that systematically exploit trust factors inherent to human interactions. The proliferation of deep falsification technologies (deepfakes), adaptive malware architectures, and the automated orchestration of multiphase attacks evidence the progressive insufficiency of control mechanisms based exclusively on static paradigms or predetermined signatures.

Parallel to its instrumentalization for offensive purposes, research has identified security vulnerabilities intrinsic to the artificial intelligence systems themselves. Contrary to initial perceptions, these systems present significant susceptibilities to manipulations executed by malicious agents. It has been determined that AI systems are subject to specific vulnerabilities such as algorithmic biases, limitations in explicability and interpretability, critical dependence on massive volumes of data, and risks derived from the compromise of underlying components in their architecture. Of particular significance, adversarial attack techniques (evasion) and systematic data poisoning emerge as priority concerns in this context, considering that their fundamental objective consists of inducing error or manipulating algorithmic models through alteration of input data or training sets, directly compromising the reliability and integrity of the systems, including those specifically designed for cybersecurity functions. This reality configures a paradox of notable relevance: the same technology conceptualized to provide protection introduces attack vectors characterized by superior levels of complexity that compromise its own operational reliability.

In the field of defensive applications of artificial intelligence, its transformative potential to counter contemporary cyber threats has been corroborated. The integration of AI significantly optimizes traditional defensive mechanisms such as Intrusion Detection Systems through automation of identification and response processes, implementation of predictive analysis, and dynamic adaptation against emerging threats. Its applications encompass a broad spectrum that includes advanced malware detection, behavioral analysis, optimized vulnerability management, and automated incident response. Nevertheless, critical evaluation evidences that the implementation of defensive architectures based on AI faces substantial challenges. Fundamentally, these systems must operate in an adversarial environment where attackers equally implement AI technologies, and, as previously documented, the defensive systems themselves based on this technology are susceptible to the same adversarial attack vectors and vulnerabilities they aim to mitigate. This circumstance configures an evolutionary dynamic characterized by technological escalation, where the adaptive capacity of hostile agents continuously challenges the effectiveness and reliability of defensive infrastructures based on artificial intelligence.

The analytical integration of these findings underscores that security management in the AI era cannot be approached through fragmentary or exclusively technical approximations. Defensive strategies must be characterized by their integrality, encompassing not only the protection of underlying ICT infrastructures and specific constitutive elements of AI systems (data, models, processes), but also considering the risks inherent to the implementation of this technology in sectors classified as high risk. Significantly, the human factor emerges not merely as a priority target of social engineering attacks enhanced by AI, but simultaneously as an indispensable defensive component in security architectures. Human resilience, strengthened through advanced awareness programs oriented to drive effective behavioral transformations, constitutes a fundamental element to complement technological defenses against attack vectors characterized by progressively superior levels of personalization and credibility. The exclusive dependence on technical controls becomes unsustainable in contexts where hostile agents manipulate psychological factors through the instrumentalization of AI technologies.

Security management requires specific adaptation to the particularities of artificial intelligence systems, necessitating the implementation of dynamic risk assessments and specialized testing methodologies that consider the continuous evolution of systems with self-learning capabilities. Algorithmic reliability constitutes a critical factor requiring continuous evaluation. The development of the regulatory framework, paradigmatically exemplified by the European Union AI Act, reflects the institutional recognition of these risks, establishing systematic categorizations based on risk levels and implementing differentiated requirements regarding security and transparency, with special emphasis on systems classified as high risk and implementations of generative AI. This normative evolution evidences the growing importance of governance in this field and the imperative need to strategically align cybersecurity mechanisms with emerging normative and ethical frameworks.

References

Carneiro Pacheco, F. (2024). Legal developments on cybersecurity and related fields. Cham: Springer Nature Switzerland.

Ciampa, M. (2024). Security awareness: applying practical cybersecurity in your world. Boston: Cengage.

Coates, C. (2022). Global cybercrime and cybersecurity laws and regulations:: issues and challenges in the 21st century. New York: Nova Science Publishers, Inc.

Cuerda Arnau, M. (2013). Nuevas amenazas a la seguridad nacional: terrorismo, criminalidad organizada y tecnologías de la información y la comunicación. Valencia: Tirant lo blanch.

Deutsch Franco, V. (2022). Ciber seguridad para directivos: [riesgos, control y eficiencia de las tecnologías de la Información]. Madrid: LID Editorial.

Elbayadi, M. (2021). Big breaches: cybersecurity lessons for everyone. New York: Apress.

Golden, G. (2025). Cybersecurity in context: technology, policy, and law. New Jersey: John Wiley & Sons, Inc, Hoboken.

Howard, R. (2023). Cybersecurity first principles: a reboot of strategy and tactics. New Jersey: John Wiley & Sons, Hoboken.

Jensen, B. (2022). Information in war: military innovation, battle networks, and the future of artificial intelligence. Washington DC: Georgetown University Press.

Kshetri, N. (2021). Cybersecurity management: an organizational and strategic approach. Toronto: University of Toronto Press.

Lawrence, S. (2024). Security in computing. Boston: Addison-Wesley Professional.

Masys, A. (2024). Navigating the Intersection of Artificial Intelligence, Security, and Ethical Governance Sentinels of Cyberspace. Cham: Springer International Publishing AG.

Moschovitis, C. (2021). Privacy, regulations, and cybersecurity: the essential business guide. Hoboken: John Wiley & Sons, Inc.

Rains, T. (2023). Cybersecurity threats, malware trends, and strategies: discover risk mitigation strategies for modern threats in your organization. Birmingham: Packt Publishing.

Seiersen, R. (2023). How to measure anything in cybersecurity risk. New Jersey: Wiley, Hoboken.
Skulmoski, G. (2022). Shields up: cybersecurity project management. New York: Business Expert Press.
ST GEORGE, S. (2024). AI in Criminal Justice Education: Assessing Educator's Attitudes and Experiences. Journal of Criminal Justice Education, 1–23.
Strang, D. (2023). Cybersecurity for decision makers. Boca Ratón: CRC Press, Taylor & Francis Group.
Thomson, L. (2024). Artificial intelligence: legal issues, policy, and practical strategies. Chigaco: ABA Publishing, American Bar Association.
Vaidya, J. (2024). Artificial intelligence security and privacy. Singapore: Springer.
Watters, P. (2024). Cybercrime and cybersecurity. Boca Raton: CRC Press, Taylor & Francis Group.

Environmental Security and Public Forces in Colombia: Integrated Strategies for the Protection of Natural Resources

Luisa Fernanda Villalba-García, Andrés Eduardo Fernández-Osorio, and Elmers Freddy Velandia-Pardo

Abstract Colombia is recognized as one of the world's most megadiverse countries, with an enviable geostrategic position. This is a strategic advantage that has been exploited by illegal actors and criminal networks affecting negatively the environment and communities. Postmodern security studies recognize environmental security as central to human development and sustainability. In this line, the present chapter argues that Colombia's public forces have assumed an increasingly central role in protecting natural resources, not only as an operational necessity in conflict zones but also as a strategic imperative within a multidimensional security framework. Colombia exemplifies how environmental protection, and national defense can converge in postmodern security frameworks through integrated operations, interinstitutional alliances, and evolving doctrine. Therefore, the country serves as an excellent case study to analyze how Public Forces have expanded their role according to the growing needs and demands that contemporary States need to address.

Keywords Colombia · Environmental security · Multidimensional security · Natural resources protection · Public forces

L. F. Villalba-García (✉) · E. F. Velandia-Pardo
Universidad Militar Nueva Granada, Bogotá, D.C, Colombia
e-mail: luisa.villalba@esmic.edu.co

E. F. Velandia-Pardo
e-mail: elmers.velandia@unimilitar.edu.co

L. F. Villalba-García
Facultad de Ciencias Militares, Escuela Militar General José María Córdova, Bogota, Colombia

A. E. Fernández-Osorio
Universidad Gabriela Mistral, Santiago, Chile
e-mail: andres.fernandez@academico.ugm.cl

L. A. Garcia-Segura (ed.), *Collaborative Insights on Technology and Legal Frameworks*, Advanced Sciences and Technologies for Security Applications,
https://doi.org/10.1007/978-3-032-27714-5_11

1 Introduction

In recent decades, the notion of security has undergone a profound transformation within academic and policy circles. Traditional security definitions, long dominated by state sovereignty and military conflict, have been challenged and expanded by more comprehensive, postmodern approaches in security studies. One significant outcome of this evolution is the rise of environmental security, which posits that ecological stability and natural resource governance are integral to the safety of nations and communities. Researchers and policymakers now examine how environmental degradation, resource scarcity, and climate change can catalyze conflict or instability, broadening the security agenda beyond its conventional bounds. This conceptual shift underscores that protecting the environment is a matter of ethical stewardship and a strategic imperative within contemporary security frameworks.

Given its unique ecological and geopolitical context, Colombia's case exemplifies these dynamics in a particularly compelling way. Renowned as one of the world's most biodiverse countries, Colombia is home to an extraordinary array of ecosystems—from Amazonian rainforests and Andean cloud forests to Pacific mangroves and Caribbean coral reefs. This natural wealth, however, comes with acute vulnerabilities: the country's abundant resources have frequently been targeted by illegal exploitation and environmental crimes, often in areas of weak governance or lingering armed conflict. Over decades of internal strife, illicit economies such as narcotics production, illegal mining, and timber trafficking have thrived in remote regions, resulting in deforestation, pollution, and biodiversity loss. These environmental challenges carry strategic significance, as they erode Colombia's ecological heritage, fuel social tensions, and threaten long-term national security.

Confronted with this convergence of ecological fragility and security concerns, the Colombian state has progressively expanded the role of its public forces—namely, the military and the national police—into environmental protection. Traditionally focused on counterinsurgency and law enforcement, these institutions in recent years have also been tasked with safeguarding national parks, combating illegal logging and mining, and dismantling trafficking networks that harm biodiversity. This broadened mandate aligns with a multidimensional security paradigm—officially endorsed in Latin America—that recognizes non-traditional threats such as environmental degradation as integral to national and human security. By integrating environmental objectives into their strategy and operations, Colombia's security forces exemplify an adaptive response to emerging threats that transcends conventional defense duties. Their involvement illustrates an evolving security paradigm in which protecting ecosystems is seen as vital to maintaining peace and stability.

In this light, Colombia offers a compelling case study of how defense institutions can adapt to non-traditional threats. The country's experience demonstrates how military, and police institutions can redefine their roles to address challenges like environmental crime and ecological degradation, which do not fit neatly into conventional security categories. An analysis of Colombia's approach offers insights into the opportunities and dilemmas of incorporating environmental protection into the core mission of public security forces. Ultimately, this case illuminates broader questions regarding the capacity of security sector institutions to evolve in response to new types of threats in the twenty-first century.

This chapter outlines the theoretical context of environmental security, tracing its conceptual evolution within security studies to establish the analytical framework. Next, the analysis turns to Colombia's environmental threats and vulnerabilities, highlighting how issues like deforestation, pollution, and illegal resource exploitation translate into security concerns. The subsequent section examines the responses of Colombia's public forces, describing how military and police institutions have incorporated environmental protection into their strategies and operations. This is followed by assessing the challenges and limitations of implementing these expanded roles, from legal and institutional hurdles to on-the-ground difficulties. Finally, the article discusses integrated strategies for safeguarding natural resources, focusing on inter-agency coordination, community engagement, and policy recommendations to reinforce the synergy between environmental conservation and security.

2 Environmental Security and Public Forces

Environmental threats are non-traditional, transnational, and multidimensional, core elements of postmodern security studies, as these do not conform to conventional military paradigms. In strategic terms, this reflects a broader trend in which the boundaries between war, peace, and crisis are increasingly blurred, requiring armed forces to adapt to non-conventional threats, including environmental degradation (Wirtz 2019). They encompass climate change, biodiversity loss, and resource scarcity, which require different security approaches. Environmental issues transcend national borders, affecting multiple countries and regions simultaneously. They impact various aspects of society, including economic stability, public health, and social cohesion, making them complex and multifaceted security concerns (Floyd 2012). Therefore, collaborative international responses are increasingly necessary (Swain et al. 2023).

Peoples and Vaughan-Williams (2021) analyze the evolution of environmental security within Critical Security Studies. The concept emerged in the late twentieth century as a response to the inadequacy of traditional military-based security definitions. It redefined security as protecting ecosystems and natural resources vital to human survival (Hough 2019). Environmental security may be seen as a factor in violent conflict, as a risk to national security, as threatened by war and military activity, or as a risk to human security and therefore a *common issue* to humanity's future (Dabelko 2022). It is also recognized that environmental stress can exacerbate social instability, migration, and conflict over scarce resources, qualifying it as a threat multiplier (Peoples and Vaughan-Williams 2021). In fact, Badie et al. (2011) state that the environment is often a source of conflict, and its threats pose long-term dangers to security.

According to Dabelko (2022), environmental security naturally implicates the role of armed forces, prompting reflection on their role in this context. Modern militaries increasingly embed environmental protection in their operational planning and training protocols, which is considered essential to national stability and human development. Environmental threats can undermine civilian safety and military interoperability; thus, pursuing environmental security is a shared responsibility between defense and civilian institutions (Katsos 2018). This adaptation reflects a broader strategic culture shift in which militaries are redefined as warfighters and as agents of governance and national resilience (Wirtz 2019). This evolving role underscores a strategic necessity of integrating environmental considerations into the core missions of public forces within a multidimensional security framework.

Consequently, preventive security strategies are necessary, including state and military adaptation to ecological risk. These risks span the land, air, and maritime domains, each presenting unique threats to civilian safety and military operability, as recognized in recent North Atlantic Treaty Organization (NATO) doctrine (Katsos 2018). Military organizations are increasingly involved in disaster response, conservation support, and pollution mitigation as part of their evolving mission sets. Such involvement is often justified by military institutions' logistical capacity and territorial reach (Peoples and Vaughan-Williams 2021). Governments have moved to integrate environmental policy into national security and defense planning, including doctrinal changes, operational guidelines, and the establishment of specialized units. However, scholars raise concerns about the militarization of environmental governance, especially when local communities are excluded or when coercive force is used to achieve conservation goals (Peoples and Vaughan-Williams 2021). In this context, environmental security is increasingly conceptualized, as Wirtz (2019) argues, as a form of strategic governance, requiring public legitimacy and interinstitutional cooperation beyond the battlefield.

3 Colombia's Strategic Landscape and Environmental Threats

Colombia is distinguished as one of the most biodiverse countries in the world. Its tropical rainforests, high-altitude Andean ecosystems, vast Amazon basin, and dual coastlines on the Pacific and Atlantic Oceans house different species. This ecological wealth is a strategic asset but also a vulnerability, as it attracts legal and illegal exploitation by domestic and foreign actors. Its strategic location also makes Colombia a critical node for maritime trade, migration routes, and transnational security issues across Central and South America.

Also, the post-conflict period following the 2016 peace accord with the guerrilla group Revolutionary Armed Forces of Colombia (FARC, by its initials in Spanish) has further complicated Colombia's environmental governance (Villamil-Castillo 2020). As regions formerly controlled by the FARC were left without centers of power and with empty spaces, these areas became even more exposed to deforestation, illegal mining, illicit crops, and transnational exploitation (Villamil-Castillo 2020). Thus, Colombia's environmental security is tied to its national security and defense strategies.

Colombia's natural resources are under constant and increasing pressure due to a complex interplay of direct and indirect threats, often exacerbated by armed conflict and institutional weakness. Deforestation is one of the most visible and serious threats—although fluctuating, annual rates remain alarming. According to the Humboldt Institute linked to the Ministry of Environment and Sustainable Development in Colombia, the most impacted areas by deforestation are the Amazon and Pacific region and the Andes in northern Antioquia (2024). The main drivers include extensive cattle ranching, the planting of illicit crops (coca), illegal logging, and illicit mining (Pirela-Ríos et al. 2023). Deforestation entails biodiversity loss, affects water regulation, increases erosion, and contributes significantly to greenhouse gas emissions.

Figure 1 displays annual deforestation trends in Colombia from 2014–2015 to 2022–2023, based on data from the Colombian Institute of Hydrology, Meteorology and Environmental Studies (IDEAM 2024). The graph shows a notable increase in deforestation between 2014 and 2015 (approx. 130,000 hectares -ha-) and a peak in 2016 and 2017 (approx. 220,000 ha). After 2016–2017, the trend shows a gradual decline, with minor fluctuations between 2018 and 2021, where deforestation levels stabilized between 160,000 and 180,000 ha. Then, a sharp drop is observed after 2021–2022, culminating in the lowest recorded level for 2022–2023, at approximately 79,000 ha.

According to the Ministry of Environment and the IDEAM (MinAmbiente & IDEAM, 2023), an estimated 79,256 ha. of forest were lost nationally, reflecting a

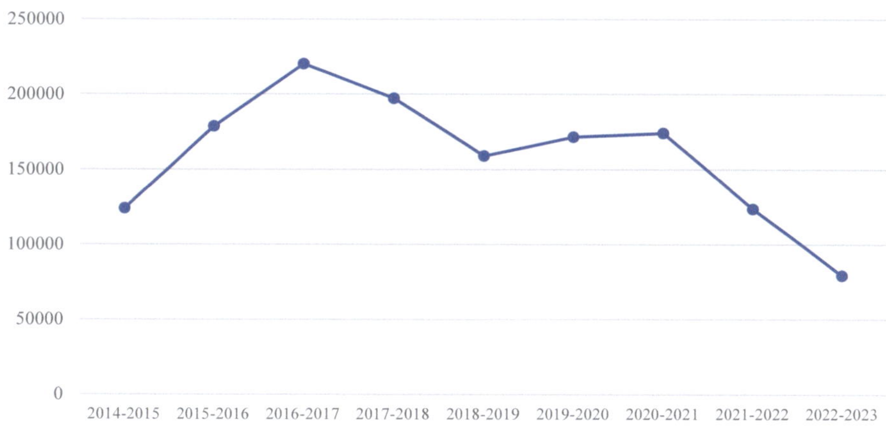

Fig. 1 Deforestation in hectares (ha). *Source* Own elaboration with data from IDEAM (2024)

continued downward trend in deforestation. This decline is associated with a deceleration pattern in key regions, including southern Meta, the Andean foothills of Putumayo, the Catatumbo region, and parts of the Pacific. Nonetheless, localized fronts of accelerated deforestation persist in areas such as Sabanas del Yarí (Caquetá), El Retorno (Guaviare), Mapiripán (Meta), the southern zone of the Serranía de San Lucas (Antioquia), and central-southern Chocó. The primary direct drivers of deforestation identified for 2023 include land clearing for speculative land grabbing, unsustainable extensive cattle ranching, unplanned infrastructure development, illicit crop cultivation, illegal mineral extraction, illegal logging, and agricultural frontier expansion into restricted zones, as well as biophysical factors such as windthrow and landslides. Notably, the monitoring system identified 533 community actors nationwide with the potential or current capacity to engage in forest conservation and community-based monitoring processes, 83 located in municipalities classified within the 2023 Active Deforestation Nuclei (NAD).

These national trends reflect broader regional dynamics and structural challenges identified in recent studies. For instance, in the Pacific region, municipalities such as Riosucio and Río Quito, in Chocó, have experienced persistent deforestation between 2015 and 2020, primarily influenced by proximity to previously cleared areas, alluvial gold mining zones, and river networks—factors projected to continue driving forest loss through 2025 without sustained local interventions (Restrepo et al. 2023). In the Western Cordillera of Antioquia, highly biodiverse areas have suffered deforestation and fragmentation rates exceeding 90%, including inside protected zones, revealing the limited effectiveness of current conservation instruments and enforcement mechanisms (Calderón-Caro and Benavides 2022). Similarly, in the Amazon region, Caquetá stands out with historically high deforestation rates, surpassing 6% forest loss between 2010 and 2018, driven by pasture expansion, palm monocultures, and illegal timber extraction, despite being subject to official conservation frameworks (García-Capdevilla et al. 2023). Moreover, policy interventions such as the National Illicit Crop Substitution Plan (PNIS), though designed to mitigate deforestation indirectly, have sometimes produced counterproductive outcomes. Delays in implementation and lack of environmental safeguards contributed to land conversion for cattle grazing and renewed coca cultivation in targeted territories (Dueñas Cavelier et al. 2023). These findings underscore the importance of advancing differentiated strategies that strengthen environmental governance while integrating ecological, socio-political, and community-based perspectives.

Alluvial gold mining (EVOA, by its initials in Spanish) can be differentiated into illicit and regulated. Illicit mining, primarily of gold, has consolidated as a significant threat to environmental security in Colombia and traditional security, as it serves as a source of financing for armed groups and criminal networks (Ayala-Sánchez 2023). Illicit mining processes use highly destructive methods such as dredging rivers and the indiscriminate use of mercury and cyanide, generating severe contamination of water sources, affecting human health (Moreno-Ramírez et al. 2024). This activity is concentrated in remote areas, often within protected areas or ethnic territories. Figure 2 shows the behavior of illicit EVOA in hectares detected through satellite imagery and field verification.

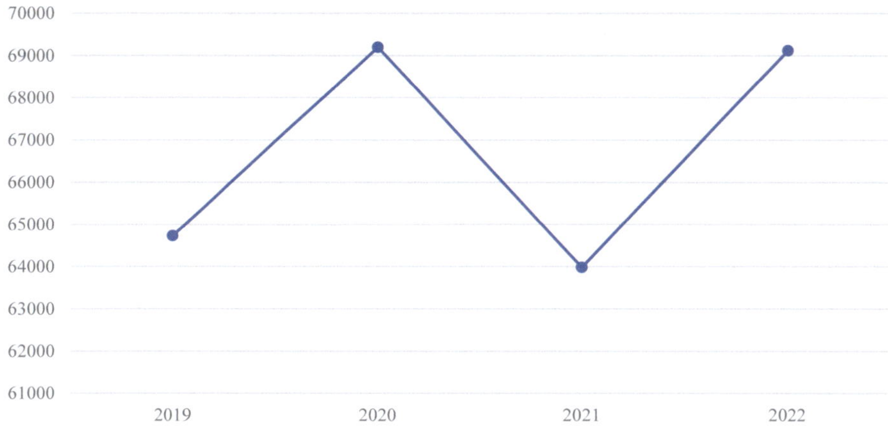

Fig. 2 Illicit EVOA in hectares (ha). *Source* Own elaboration with data from UNODC (2020, 2021, 2022, 2023)

As can be noticed, total illicit EVOA peaked in 2020 at 100,752 ha, then declined to 94,733 ha by 2022. According to the United Nations On Drugs and Crime (UNODC) reports, illicit mining remains present in the Colombian Departments Antioquia, Bolivar, Cauca, Choco, Cordoba, Nariño, Putumayo, and Valle del Cauca, with the top three departments (Antioquia, Chocó, Bolívar) consistently accounting for over 85% of EVOA. According to the reports (UNODC 2020, 2021, 2022, 2023), these regions coincide with areas of armed group presence and weak institutional control, facilitating illegal operations. It can also be evidence that, despite minor year-to-year variation, the extent of the phenomenon has remained structurally high since 2019, indicating a persistent and systemic issue. The 2022 figure of 94,733 ha is only 6% lower than the 2020 peak, suggesting a limited long-term impact of state interventions.

EVOA contributes to mercury pollution, river sedimentation, and deforestation, particularly in biodiversity-rich zones like the Atrato and Nechí river basins. The illegal economy around gold mining is closely linked to armed financing and territorial control, representing a strategic security challenge. Cortés Espitia and Cabezas Palacios (2022) confirm that 60–82% of mining in departments like Chocó and Antioquia is illegal, as well there is a strong involvement of groups such as the ELN, Clan del Golfo, and FARC dissidents, especially in Cauca, Bajo Cauca Antioqueño, and the Atrato region. These groups profit from illicit mining through extortion, protection rackets, or direct operation; thus, converging different threats to human and environmental security.

Beyond the environmental and territorial consequences, recent literature reveals that illegal mining in Colombia constitutes a multifaceted threat that merges environmental degradation with criminal governance. According to Juárez (2016), illegal mining accounts for 50–80% of total extractive activity in the country and extends to 65% of the territory, reflecting its embeddedness in local economies and its resilience to formalization efforts. Overlapping narratives sustain this embeddedness: while the legal framework defines illegal mining as a criminal act without a valid title for many local miners, it represents a traditional livelihood deeply rooted in community and territory (Juárez 2016). As noted by Sandoval and Arciniegas (2022), the activity increasingly serves as a strategic economic engine for armed groups (GAO), who have shifted toward illegal gold extraction as a more profitable and less surveilled alternative to narcotrafficking. These dynamics intensify environmental destruction through mercury contamination, deforestation, and biodiversity loss, and weaken state sovereignty by enabling alternative governance systems. Criminal organizations exploit these conditions to establish control over territories, forming enclaves where illegal mining is sustained by logistical, financial, and social regulation mechanisms (Rivera-Rhon and Bravo-Grijalva 2023). These enclaves function under a neo-extractivist model that blurs the lines between formal and informal economies, whose operation is predicated on state absence, systemic corruption, and social complicity. Thus, addressing illegal mining in Colombia transcends environmental regulation and requires a multidimensional strategy combining environmental protection with institutional reconstruction, territorial governance, and inclusive economic alternatives.

Illicit crops such as coca cultivation and, to a lesser extent, poppy cultivation continue threat the environment in Colombian territory. For instance, although the relationship between deforestation and illicit crops is not always direct, opening areas for illicit cultivation often facilitates subsequent colonization and deforestation for other purposes (Álvarez Calderón and Rodríguez Beltrán 2018). Furthermore, agrochemicals and chemical precursors contaminate soil and water, and eradication strategies (particularly aerial spraying, although currently suspended) have generated controversy due to environmental and social impacts (Rodríguez Ruiz 2021). This trend can be visualized in Fig. 3, which shows the sustained increase in coca cultivation over the past decade, based on official data from UNODC.

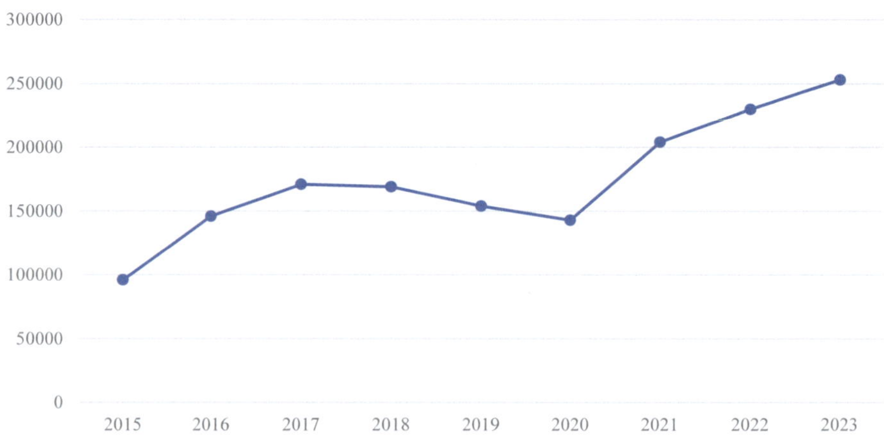

Fig. 3 Coca cultivation in hectares (ha). *Source* Own elaboration with data from UNODC (2024)

Three periods can be evidenced regarding coca cultivation in Colombia from 2015 to 2023, a rapid increase from 2016 to 2017, a stabilization with a soft decrease from 2018 to 2020, and a recent spike from 2021 to 2023. Following the Peace Agreements, a rapid expansion was fueled by opportunistic behavior by growers and new criminal actors (Acero Vargas et al. 2019). Farmers responded to the incentives (or perceived future incentives) of the peace deal's programs, while the retreat of FARC control and a temporary lull in suppression opened space for coca to spread. The increase was mainly in long-established coca zones, intensifying an already entrenched issue (Villamil-Castillo 2020). Then, starting on 2018, there was period of constraint and slight decline in coca crops, driven by robust government intervention via substitution aid and forced eradication (Torrado 2021) However, the overall coca economy remained resilient, concentrated in core areas and backed by emerging armed groups.

The reduction, while real, was modest and fragile, coca cultivation stayed near historically high levels even as it dipped from the absolute peak. By 2021, a strong rebound to new record highs, underpinned by structural market forces and the undercutting of prior enforcement gains, international demand and lucrative cocaine prices met improved cultivation and production methods, leading to greater yields and profits that incentivized growers to plant more (Posada 2022). At the same time, fragmented armed groups firmly entrenched themselves in drug-producing regions, actively promoting coca planting as a source of revenue. The state's strategies struggled to adapt, and a shift toward long-term solutions translated into a short-term relaxation of pressure on coca cultivation (Acosta 2024). Consequently, coca spread broadly and reached unprecedented levels, especially in Colombia's vulnerable peripheries.

Other threats against environmental security in Colombia are shown in Table 1. The table provides a comprehensive overview of environmental security threats in Colombia, categorized by type of threat. For each threat, it identifies the key drivers, which the underlying causes or systemic pressures contributing to the threat (e.g., global demand, deforestation, climate change, weak governance, etc.); the principal actors involved, including both legal and illegal, such as industries, state entities, armed groups, local mafias, foreign fleets, and even international networks; and the effects on environmental security with specific consequences of each threat, including biodiversity loss, pollution, water scarcity, forest degradation, and weakened sovereignty.

The main threats listed include wildlife trafficking, water pollution, climate vulnerability, illegal mining, coca cultivation, IUU fishing, and geopolitical exploitation. The table evidence an overall picture that reveals how environmental degradation in Colombia is deeply entangled with organized crime, institutional weakness, and both domestic and transnational drivers, ultimately posing complex security challenges. For instance, trafficking of fauna and flora severely threatens Colombia's biodiversity, particularly in highly sensitive ecosystems such as the Amazon and Chocó. Wildlife trafficking contributes to ecological imbalance, species extinction, and disruption of local food chains (Brombacher et al. 2021). Water pollution, which is driven by illegal gold mining, contributes to releasing 205 metric tons of mercury

Table 1 Other threats to environmental security in colombia

Threat	Key drivers	Main actors involved	Effects on environmental security
Trafficking of fauna and flora	National/international demand (pets, fur, meat, timber), criminal networks, weak control/ sanctions	Local and international trafficking networks, illegal hunters/ gatherers, and middlemen	Loss of biodiversity, risk of species extinction, and ecological imbalances affect the entire territory, especially the Amazon, Chocó, Andes, and Caribbean
Water pollution	Untreated industrial/ domestic waste, agrochemicals, legal/ illegal mining, and oil spills	Industries, municipalities, the agricultural sector, mining companies, and GAO (pipeline attackers)	Human health impacts, loss of aquatic biodiversity, eutrophication, and reduced drinking water availability affect Magdalena, Cauca, and urban-industrial zones
Vulnerability to climate change	GHG emissions, deforestation, ecosystem degradation, and overdependence on water	Energy, transport, agriculture, and land use sectors	Natural disasters, water stress, food insecurity, glacial retreat, sea level rise, drought/flood risks in the Caribbean, Orinoquia, and the Andes
IUU fishing	Lack of monitoring, foreign intrusions	Foreign fleets (e.g., Nicaragua), illegal fishers	Biodiversity loss, fish stock depletion, and sovereignty threats in the Seaflower Reserve
Foreign environmental exploitation	Geopolitical disputes, extractive interests	Nicaragua (hydrocarbon auctions), Chinese firms in indigenous zones	Environmental degradation, Exclusive Economic Zones violations, and indigenous rights conflicts
Climate change impacts and ecological fragility	Global warming, land use change, and precipitation shifts	State agencies, urban planners, and infrastructure sectors	Water rationing (Bogotá), coastal erosion (Palomino), glacial retreat, food and water insecurity
Environmental defenders and governance gaps	Illegal economies, resource conflicts, and institutional fragility	Armed groups, local mafias, and corrupt officials	Targeted killings, threats to park rangers and indigenous leaders, poor law enforcement, and civil-military coordination

Source Own elaboration

annually in Colombia. This contaminates rivers, threatens aquatic life, and affects human health, especially in rural communities (Morales 2017). Also, according to Morales (2017), coca processing also pollutes water due to chemical precursors and is further exacerbated by glyphosate spraying, which has been associated with toxic runoff and health.

On the other hand, weak governance of ecosystems and accelerated deforestation (124,000 ha in 2015) increase Colombia's exposure to drought, glacial retreat, and water insecurity (Morales 2017). Expanding coca and mining into protected or remote areas also weakens ecosystem resilience. These illicit activities are not isolated—they reinforce climate vulnerabilities by undermining forest carbon sinks (Morales 2017). Also, it can be found external actors exploiting legal ambiguity or weak governance in conflict or post-conflict zones—e.g., in Orinoquia, foreign companies with government backing have pressured land use shifts that violate constitutional protections for peasant landowners and environmental reserves (Morales 2017), These trends reveal the blurred boundaries between legal and illicit (Brombacher et al. 2021), and the failure of state policy to prioritize environmental integrity in the face of economic pressure.

Also, Colombia has faced the consequences of IUU fishing in its extensive Exclusive Economic Zone, particularly in the Caribbean and Eastern Tropical Pacific, as a zone highly vulnerable due to distant water fleets, including subsidized foreign actors like China (Montecalvo et al. 2023). IUU fishing undermines conservation zones such as the Seaflower Biosphere Reserve and Malpelo Sanctuary (Rodríguez Ortiz et al. 2016), threatening endangered marine species of sea turtles and marine fish. As with the crimes explored before, IUU fishing in Colombia often overlaps with narco-trafficking, arms smuggling, and human trafficking, revealing a pattern of environmental crime convergence that complicates enforcement and multiplies risks to maritime security (ELI and Jjay 2023). Also, IUU activity directly impacts artisanal fishers and coastal communities, especially in San Andrés y Providencia and Pacific departments, eroding fish stocks and exacerbating food insecurity and social vulnerability.

Finally, agreeing with Van Uhm (2023), organized crime is linked to environmental threats. Van Uhm (2023), exposes how illicit economies are not isolated but function as complex, overlapping systems. Environmental threats often coexist in regions with low governance capacity, operating through shared infrastructure, labor markets, and protection mechanisms, including corruption and violence. Thus, environmental degradation and organized crime are co-dependent. These crimes finance armed conflict and accelerate ecological collapse, particularly in biodiversity hotspots.

4 Strategic Input from Public Forces to Environmental Security

As explored above, in the Colombian context, environmental security has taken on urgency as environmental degradation becomes entwined with violent conflict and organized crime. The rise in these crimes not only endangers Colombia's biodiversity but also threatens to destabilize the fragile peace in former conflict regions (Delgado and Edle von Peter 2024). A defining feature of Colombia's environmental security landscape is the intersection of ecological degradation with the presence of armed groups and criminal networks. Some rebel fronts paradoxically limited deforestation during the conflict by controlling access to remote forests (Delgado and Edle von Peter 2024). With their withdrawal, empty spaces (Álvarez-Calderón 2017), that could also be called security vacuums emerged, quickly exploited by criminal organizations and dissident armed factions for illegal resource extraction

Colombia's constitutional and legal framework establishes a clear responsibility for the State—including its public forces—in protecting the environment. The Constitution (República de Colombia 1991), environmental protection as a collective right and a duty. Article 8 declares that it is an obligation of the state and individuals to protect the nation's natural wealth, and Article 79 further guarantees everyone the right to a healthy environment and mandates the State to protect biodiversity and ecological integrity. In addition, Article 95(8) imposes on all citizens the duty to "protect the country's cultural and natural resources and to ensure a healthy environment" (República de Colombia 1991). These provisions lay the groundwork for involving all state institutions—including the military and police—in environmental stewardship. Consequently, the Constitution frames environmental protection as integral to the general welfare and part of Colombia's legal order, thereby justifying state action (even by security forces) to prevent environmental degradation that could infringe on constitutional rights.

Building on these constitutional principles, Colombia enacted Law 99 (Congreso de la República de Colombia 1993), a landmark environmental law that created the Ministry of Environment and the National Environmental System (SINA by its initials in Spanish). Law 99 restructured environmental governance by decentralizing authority to regional agencies and establishing mechanisms for citizen participation in environmental management. Crucially, it also delineated roles for other branches of government. Article 103, explicitly assigns the National Ministry of Defense and the Armed Forces a role in environmental protection, stating that the Armed Forces must "ensure the protection and defense of the environment and renewable natural resources and enforce the norms issued to protect the natural patrimony of the nation as an element of national sovereignty" (Acuña-Montañez 2023). This provision embeds environmental duties into the mission of the military. In essence, safeguarding Colombia's environmental patrimony is viewed as part of defending the nation—a recognition that environmental integrity underpins long-term security and sovereignty.

Colombian National Police, part of the Ministry of Defense, also derive environmental responsibilities from law and policy. While the police are a civilian force charged primarily with maintaining public order according to the article 218 of the Constitution (República de Colombia 1991), environmental legislation confers specific duties to them. Law 99 of 1993 granted “police power” to environmental authorities, meaning that environmental agencies can call upon the National Police to enforce regulations and sanctions (Blackman et al. 2005). Over time, the National Police have developed specialized units to fulfill this mandate.

The National Directorate of Carabineros and Environmental Protection (DICAR) is a police branch dedicated to rural security and environmental law enforcement. Through DICAR, police officers engage in protecting natural parks, controlling wildlife trafficking, and supporting investigations of environmental crimes in coordination with environmental agencies and prosecutors. In recent years, Colombia has also strengthened its legal arsenal. For example, the Environmental Crimes Law (Congreso de la República de Colombia 2021) was passed to criminalize deforestation explicitly and create five new environmental offenses (such as financing of illegal mining and wildlife trafficking), with penalties of up to 15 years in prison. This evolution in the legal framework—from the Constitution and Law 99 of 1993 to new criminal statutes—underpins a growing expectation that the military and police (collectively the public forces) will actively contribute to environmental protection alongside civilian environmental institutions.

Colombian National Army has been at the forefront of Colombia’s militarized campaigns against environmental crimes and participating in civic action projects to restore ecosystems. A flagship initiative was Campaña Artemisa (Campaign Artemis), launched in 2019 as a major offensive against deforestation. This campaign mobilized joint military, police, and environmental personnel (roughly 17,000 troops in initial phases) to target illegal logging, land clearing, and other environmental offenses in critical areas like national parks (Dussán 2019). Under Artemisa, Army units—often alongside National Police and park rangers—conducted raids in deforestation hotspots, removed illegal settlers and miners, and destroyed illicit coca processing labs that were causing ecological damage. According to Dussan (2019) early operations focused on prized ecosystems such as Chiribiquete National Park and the Amazon rainforest frontiers, where organized crime groups had been burning and clearing forest for cattle and crops. The author points out that the Army leveraged its logistical capabilities and intelligence resources (including surveillance drones) to locate environmental offenders. This has also contributed to territorial control—deterring illegal land seizures and providing security for environmental officials. The Army’s deep familiarity with Colombia’s rugged geography and its permanent deployment in rural regions give it a unique capacity to identify areas requiring protection and respond rapidly (Acuña-Montañez 2023). Army commanders emphasize that these capabilities, traditionally used in counterinsurgency, are now being repurposed to defend ecological corridors and natural reserves as part of the Army’s evolving mission.

The Colombian National Navy also plays a crucial role in guarding Colombia’s extensive maritime and riverine environments, from the Pacific and Caribbean coasts to the Amazon and Orinoco rivers. A key contribution is maintaining maritime domain awareness to combat illegal fishing, pollution, and trafficking threatening marine biodiversity (Armada Nacional de Colombia 2022). The Navy's coast guard units conduct continuous patrols around marine protected areas such as the Seaflower Biosphere Reserve in the San Andrés archipelago, where they have intercepted numerous illegal fishing operations. In a 2024 operation, for example, the Navy detained 49 individuals engaged in illegal fishing around Quitasueño (a highly protected atoll), seizing 182 kg of illegally caught lobsters and crab and confiscating boats and scuba equipment used in poaching (Romero 2024). Such actions aim to counter the environmental scourge of unsustainable fishing and protect endangered marine species in Colombian waters.

The Navy also supports the enforcement of bans on shark finning and sea turtle hunting by monitoring ports and shipping routes for contraband wildlife products. In addition, the Colombian Navy has assumed a leadership role in Campaign Orion, a multinational naval campaign against narcotrafficking (COGFM 2022). While primarily a counternarcotics effort, Orion yields environmental benefits by interdicting cocaine shipments and the supply chains behind them (Rivera-Páez 2022). Since 2018, Orion's collaborative operations (involving over 40 nations) have seized more than 545 tons of cocaine and 224 tons of marijuana (Reuters 2021). This has a direct environmental impact, as to produce such quantities, hundreds of thousands of hectares of rainforest would likely have been cut down for coca cultivation and drug labs. Thus, by disrupting narco-networks (which drive deforestation and chemical pollution), the Navy contributes to environmental security.

The Colombian Air Force has extended the reach of environmental protection through aerial surveillance, rapid response to forest fires, and disaster relief missions. With Colombia's vast and often inaccessible terrain, the Air Force's intelligence, surveillance, and reconnaissance (ISR) assets are invaluable for detecting environmental threats. High-altitude aircraft and drones operated by the Air Force (and Army aviation) have been used to map deforestation in real time, spot illegal mining camps, and monitor protected areas for encroachment (Fuerza Aérea Colombiana 2011). This aerial perspective feeds into multi-agency efforts like Operation Artemisa, where Air Force intelligence helped pinpoint illegal clearings in national parks (Dussán 2019).

In addition to surveillance, the Air Force has become a key player in wildfire response, an increasingly critical task as climate change and land burning practices lead to more frequent forest fires. Every dry season, the Air Force works jointly with Colombia's Disaster Risk Management Unit, firefighters, civil defense volunteers, the Army and National Police to combat wildfires across the country (Estrada Villa and Jaramillo García 2019). The Air Force operates specialized equipment such as Bambi Bucket water tanks slung under UH-60 Black Hawk and Huey helicopters, capable of dropping hundreds of gallons of water or fire retardant onto active fires (Diálogo Américas 2015). The Air Force has deployed helicopters from its Combat Air Commands to douse large wildfire coordinating closely with ground firefighters to save local watersheds.

The Air Force's airlift capabilities also allow it to deliver humanitarian assistance in environmental emergencies—whether transporting firefighters and supplies to remote wildfire sites, or evacuating communities during floods, landslides, and extreme weather events exacerbated by environmental degradation (Jaimes Gómez 2021). Through these efforts, the Air Force enhances Colombia's preparedness and resilience to environmental disasters, complementing its security-oriented missions with vital environmental protection services.

Despite the array of initiatives by military and police forces, Colombia faces significant challenges in coordinating and integrating these efforts with broader environmental governance. Addressing environmental challenges require a shift toward more integrated and preventive strategies that unify command structures, fuse intelligence, and place communities at the heart of environmental security efforts. Colombian authorities have signaled the start of such a shift. Recently, the Minister of Environment announced that the military and police will refocus on *preventive* measures and work hand-in-hand with local communities to contain deforestation (Camargo Páez 2024). This reflects an important strategic reorientation: rather than relying on punitive raids alone, the public forces will emphasize early warning and partnership.

Enhanced civil-military cooperation is equally crucial. The public forces should act not in isolation but in concert with civilian institutions and the communities that are the environment's long-term guardians (Molina-Orjuela et al. 2022). This means incorporating civil authorities and local leaders into the operational planning process. For example, before conducting an operation in a national park, the joint command would coordinate with National Parks officials and Indigenous or campesino community representatives in that area. Such cooperation can ensure that operations are sensitive to local social dynamics and that there is follow-up in the form of community support afterward.

Colombia's peace accords and rural reforms offer a framework for this, the 2016 Peace Agreement emphasized integrating conflict-affected regions through development and local governance (El Gobierno Nacional y las FARC-EP 2016). Environmental security strategies should dovetail with these peacebuilding measures. As Camargo-Paez (2024) argued, the solution to environmental harm "cannot be to regard members of marginalized communities as criminal actors" but to actively involve those communities in safeguarding forests. In practical terms, this could involve training and employing local inhabitants as park rangers, forest monitors, or eco-guides, providing them with alternative livelihoods in conservation. The Armed Forces can support such initiatives by offering logistics for remote conservation projects or helping build infrastructure (like ranger stations or eco-tourism facilities) reinforcing a benign state presence. Meanwhile, the police can expand community policing programs that encourage residents to report illegal logging and mining, with guarantees for their safety.

A preventative approach also entails a stronger focus on alternative development and environmental governance reform. Since poverty and lack of economic options drive much of the illicit exploitation, the government (with support from public forces in a security role) should increase efforts to provide alternatives to communities. This includes accelerating coca crop substitution programs, sustainable agro-forestry projects, and land formalization in areas of high deforestation. The pressure on ecosystems can be reduced by offering farmers viable livelihoods that do not require clearing more forest. For instance, supporting agroforestry or ecotourism in Amazon regions can improve incomes and incentivize forest conservation. The public forces can facilitate these programs by securing the areas and building trust so that government agronomists and educators can work there safely.

Environmental governance must be strengthened: the capacities of civilian environmental institutions need bolstering so they can take the lead, with the military and police in a supporting role. This may involve anti-corruption measures to ensure that environmental licenses are not misused, and illegal loggers cannot bribe their way out of sanctions. It also means improving land-use planning – for example, fully implementing territorial environmental zoning to demarcate and respect protected areas and ecological corridors in development plans.

Finally, community participation should be institutionalized. Mechanisms like environmental councils or local vigilance committees could be created, where community members, local officials, and security representatives regularly meet to discuss emerging environmental threats and jointly devise solutions. Such participatory forums align with the ethos of SINA and the rights of ethnic communities, as recognized in Colombian law. They help ensure that enforcement is balanced with social justice. When communities are treated as partners, not targets, they become force multipliers for environmental security. Indeed, successful conservation in conflict-prone areas often relies on local knowledge–residents can warn of illicit activity and mediate conflicts over resources. The armed forces and police, in turn, can provide the backbone of security and technical assistance, while respecting human rights and local autonomy.

By combining the strength and reach of public forces with the knowledge and engagement of communities and environmental entities, Colombia can move from reactive militarized campaigns to sustainable environmental security. This integrated model promises to protect the nation's invaluable natural endowment while consolidating peace and development in its most vulnerable territories. Every tree saved and river kept clean is not only an environmental victory but also a step toward a more secure Colombia, achieved through the strategic synergy of public forces and society itself.

5 Conclusions

Environmental security in Colombia is deeply intertwined with organized crime, armed conflict, and institutional fragility, making it a core concern for national defense and public security. Therefore, Colombian Public Forces (Army, Navy, Air Force, and Police) have adapted their missions to address environmental threats, moving beyond traditional roles into conservation, enforcement, surveillance, and disaster response.

Illegal economies like mining, coca cultivation, and IUU fishing function as strategic threats to environmental and national security, financing armed groups and eroding state sovereignty. Yet, in this context, Constitutional and legal frameworks provide clear mandates for military and police involvement in environmental protection, reinforcing the legitimacy of their interventions. As a result, key operations such as Campaign Artemisa and Operation Orion illustrate the operational capabilities of the public forces but also highlight limitations in achieving long-term environmental outcomes without structural reforms.

Coordination remains a major institutional challenge, with fragmented mandates between security forces and civilian environmental agencies undermining sustained success. Militarized approaches risk alienating local communities, especially when enforcement actions are perceived as heavy-handed or disconnected from social realities. Hence, an integrated environmental security strategy must be preventive and participatory, emphasizing intelligence sharing, civil-military coordination, and community inclusion.

This paper concludes that Public Forces should complement, not replace, civilian environmental institutions, focusing on deterrence, logistics, and protection of vulnerable areas while enabling local governance and conservation efforts. Also, intelligence fusion centers and unified command structures are urgently needed to enhance strategic coordination and prevent environmental crimes before they occur. Alternative livelihoods and territorial governance reforms are essential to reduce dependency on illicit economies and foster local stewardship of natural resources. Finally, it has been evidenced that Colombia's experience serves as a valuable case study globally, demonstrating both the opportunities and risks of integrating Public Forces into environmental security missions in post-conflict and high-biodiversity contexts.

References

Acero Vargas, C., Parada Hernández, M. M., & Machuca Pérez, D. X. (2019). La paz narcotizada: Decisiones, presiones y diseño institucional del Programa Nacional Integral de Sustitución de Cultivos de Uso Ilícito. *Análisis Político, 32*(97), 114–135. https://doi.org/10.15446/anpol.v32n97.87195

Acosta, L. J. (2024). *Colombian coca leaf farming hit two-decade high in 2023, UN says | Reuters*. Reuters. https://www.reuters.com/world/americas/colombian-coca-leaf-farming-hit-two-decade-high-2023-un-says-2024-10-18/

Acuña-Montañez, N. M. (2023). El deber de protección del medioambiente del Ejército nacional. Un análisis desde las acciones populares. In P. A. Velásquez-Cardona & C. H. Prieto-Fetiva (Eds.), *El Ejército Nacional de Colombia y el deber jurídico de protección del medioambiente* (pp. 79–99). Sello Editorial ESMIC. https://doi.org/10.21830/9786289544664

Álvarez Calderón, C. E., & Rodríguez Beltrán, C. A. (2018). Ecosistemas criminales: Hábitats para la convergencia y la globalización desviada. *Revista Científica General José María Córdova, 16*(24), 1–30. https://doi.org/10.21830/19006586.352

Álvarez-Calderón, C. E. (2017). Ocupación de los Espacios Vacíos: Una Condición sine qua non para la seguridad multidimensional en Colombia. In C. Álvarez (Ed.), *Escenarios y Desafíos para la Seguridad Multidimensional en Colombia* (pp. 307–385). Sello Editorial ESDEG. https://esdeguelibros.edu.co/index.php/editorial/catalog/view/27/23/313

Armada Nacional de Colombia. (2022). *Sistema de Gestión Ambiental*. https://www.armada.mil.co/es/content/sistema-gestion-ambiental

Ayala-Sánchez, F. (2023). Las Fuerzas Militares y su lucha contra el agotamiento de recursos naturales como activos en Colombia. In E. A. Salamanca-Rodríguez & J. A. Serpa-Hernández (Eds.), *Seguridad humana y construcción de patria en defensa de la vida: Seguridad sanitaria, ambiental y personal: Vol. II* (pp. 455–485). Sello Editorial ESDEG. https://doi.org/10.25062/9786287602595.17

Badie, B., Berg-Schlosser, D., & Morlino, L. (2011). Environmental Security Studies. In *International Encyclopedia of Political Science*. SAGE Publications. https://doi.org/10.4135/9781412959636.n188

Blackman, A., Hoffmann, S., Morgenstern, R., & Topping, E. (2005). *Assessment of Colombia's National Environmental System (SINA)*. Resources for the future. https://www.rff.org/publications/reports/assessment-of-colombia039s-national-environmental-system-sina/

Brombacher, D., Garzón, J. C., & Vélez, M. A. (2021). Introduction Special Issue: Environmental Impacts of Illicit Economies. *Journal of Illicit Economies and Development*, *3*(1). https://doi.org/10.31389/jied.107

Calderón-Caro, J., & Benavides, A. M. (2022). Deforestación y fragmentación en las áreas más biodiversas de la Cordillera Occidental de Antioquia (Colombia). *Biota Colombiana*, *23*(1), 1–15. https://doi.org/10.21068/2539200X.942

Camargo Páez, P. A. (2024). *Artemisa: The security model of environmental defense and the rights of rural communities* |. Open Global Rights. https://www.openglobalrights.org/security-environmental-defense-rural-communities-artemisa/

COGFM. (2022). *Orion Naval Campaign is the cornerstone for the global fight against drug trafficking | Comando General de las Fuerzas Militares de Colombia*. https://www.cgfm.mil.co/en/multimedia/noticias/orion-naval-campaign-cornerstone-global-fight-against-drug-trafficking

Congreso de la República de Colombia. (1993). *Ley 99*. https://www.funcionpublica.gov.co/eva/gestornormativo/norma.php?i=297

Congreso de la República de Colombia. (2021). *Ley 2111*. https://www.funcionpublica.gov.co/eva/gestornormativo/norma.php?i=167988

Cortés Espitia, I., & Cabezas Palacios, J. V. (2022). *Aparatos organizados del poder en la economía del oro*. INDEPAZ. https://tinyurl.com/y233w447

Dabelko, G. (2022). Environmental Security. In A. Collins (Ed.), *Contemporary Security Studies* (6th ed., pp. 247–262). Oxford University Press.

Delgado, C., & Edle von Peter, T. (2024). *Demilitarizing environmental protection will benefit Colombia's biodiversity*. SIPRI. https://www.sipri.org/commentary/blog/2024/demilitarizing-environmental-protection-will-benefit-colombias-biodiversity

Diálogo Américas. (2015, February 24). Colombian Air Force Helps Fight Forest Fires. *Diálogo Américas*. https://dialogo-americas.com/articles/colombian-air-force-helps-fight-forest-fires/

Dueñas Cavelier, Carolina, García Rodríguez, Nicolás, & Mejía Gil, Nicolás Felipe. (2023). Consecuencias no intencionadas de la política de drogas: Evidencia del PNIS sobre la deforestación en Colombia. *CESED*. https://doi.org/10.57784/1992/69591

Dussán, Y. (2019, September 20). Colombia Launches Military Campaign Against Deforestation. *Diálogo Américas*. https://dialogo-americas.com/articles/colombia-launches-military-campaign-against-deforestation/

El Gobierno Nacional y las FARC-EP. (2016). *Acuerdo Final para la Terminación del Conflicto y la Construcción de una Paz Estable y Duradera*.

ELI, & JJAY. (2023, June). *Environmental Crime Convergence: Launching an Environmental Crime Convergence Paradigm Through Investigation of Transnational Organized Crime Operations*. https://tinyurl.com/44xn92xn

Estrada Villa, E. J., & Jaramillo García, W. A. (2019). Editorial Revista Ciencia y Poder Aéreo. *Revista Ciencia y Poder Aéreo*, *14*(1), 11–13.

Floyd, R. (2012). Climate Change, Environmental Security Studies and the Morality of Climate Security. *E-International Relations*. https://www.e-ir.info/2012/01/20/climate-change-environmental-security-studies-and-the-morality-of-climate-security/

Fuerza Aérea Colombiana. (2011). *Plan Estratégico Institucional Fuerza Aérea Colombiana: 2010–2030*. https://tinyurl.com/f5666kc4

García-Capdevilla, D. A., Buitrago-Bermúdez, O., & Aguirre, M. A. (2023). Alternativas Comunitarias a los Procesos de Deforestación en la Amazonía Colombiana. Caso el Caraño, Caquetá. *Historia Ambiental Latinoamericana y Caribeña (HALAC)*, *13*(1), 19–52. https://doi.org/10.32991/2237-2717.2023v13i1.p19-52

Hough, P. (2019). Back to the future: Environmental security in nineteenth century global politics. *Global Security: Health, Science and Policy*, *4*(1), 1–13. https://doi.org/10.1080/23779497.2019.1663128

Humboldt Institute. (2024, April 16). *Reporte Bio: Nueva radiografía del estado de la biodiversidad en Colombia*. https://www.humboldt.org.co/noticias/reporte-bio-nueva-radiografia-del-estado-de-la-biodiversidad-en-colombia

IDEAM. (2024). *Bosque y Deforestación en Cifras* |. https://www.ideam.gov.co/temas/monitoreo-de-bosques/geovisor

Jaimes Gómez, E. H. (2021). Capacidades de las Fuerzas Militares de Colombia en escenarios de asistencia humanitaria y respuesta a desastres. *Estudios En Seguridad y Defensa*, *16*(32), 319–342. https://doi.org/10.25062/1900-8325.317

Juárez, F. (2016). La minería ilegal en Colombia: Un conflicto de narrativas. *El Ágora U.S.B.*, *16*(1), 135–146.

Katsos, G. E. (2018). *The U.S. Government's Approach to Environmental Security: Focus on Campaign Activities*. National Defense University Press. https://tinyurl.com/27dv9x43

MinAmbiente, & IDEAM. (2023). *Monitoreo de la superficie de bosque y la deforestación en Colombia*. https://tinyurl.com/5969ah8j

Molina-Orjuela, D. E., Chavarro-Ospina, S. G., & Guzmán-Alvarado, B. O. (2022). Impactos del conflicto armado colombiano sobre el medio ambiente y acciones para su efectiva reparación. *Revista Científica General José María Córdova*, *20*(40), 1087–1103. https://doi.org/10.21830/19006586.1129

Montecalvo, I., Le Billon, P., Arsenault, C., & Schvartzman, M. (2023). Ocean predators: Squids, Chinese fleets and the geopolitics of high seas fishing. *Marine Policy*, *152*, 105584. https://doi.org/10.1016/j.marpol.2023.105584

Morales, L. (2017). *Peace and environmental protection in Colombia: Proposals for Sustainable Rural Development*. The Dialogue. Leadership for the Americas. https://tinyurl.com/y9ayfrta

Moreno-Ramírez, A., Paz-Moreno, D., Ledesma-Copete, R. E., & Becerra-Torres, Y. E. (2024). La Formalización Minera como Catalizador de Gobernanza Territorial en la Lucha Contra la Minería Ilegal en Colombia. *Revista Científica de Salud y Desarrollo Humano*, *5*(3), 1158–1182. https://doi.org/10.61368/r.s.d.h.v5i3.324

Peoples, C., & Vaughan-Williams, N. (2021). Environmental and planetary security. In *Critical Security Studies* (3rd ed., pp. 133–148). Routledge.

Pirela-Ríos, A., Junca, J. J., Triana-Ángel, N., & Burkart, S. (2023). *¿Qué pasa con la deforestación cuando 'acaba' el conflicto?* CGIAR. https://cgspace.cgiar.org/bitstreams/e0434f33-0e14-4be5-b63c-874cc1bc83b1/download

Posada, J. D. (2022). *Colombia Sees Historic Levels of Coca Cultivation and Cocaine Production*. InSight Crime. http://insightcrime.org/news/colombia-hits-historic-levels-cocaine-cultivation-production-unodc/

República de Colombia. (1991). *Constitución Política de Colombia*. Panamericana.

Restrepo, J., Hoyos, N., & Restrepo, J. (2023). Dinámicas espacio-temporales de la deforestación en los municipios de Riosucio y Río Quito, Chocó, Colombia. *Revista de La Academia Colombiana de Ciencias Exactas, Físicas y Naturales*, *47*(185), 977–985. https://doi.org/10.18257/raccefyn.1961

Reuters. (2021). Global coalition seizes 145.3 tonnes of cocaine. *Reuters*. https://www.reuters.com/world/americas/global-coalition-seizes-1453-tonnes-cocaine-colombias-navy-2021-11-29/

Rivera-Páez, S. (2022). Campaña Naval y Fluvial Orión: Una oportunidad de Colombia para fortalecer la cooperación con la OTAN. *Revista de Las Fuerzas Armadas, 259*, 67–80. https://doi.org/10.25062/0120-0631.397

Rivera-Rhon, R. A., & Bravo-Grijalva, C. E. (2023). Gobernanzas criminales y enclaves productivos de la minería ilegal en Ecuador. *Revista Logos Ciencia & Tecnología, 15*(2), 49–69. https://doi.org/10.22335/rlct.v15i2.1734

Rodríguez Ortiz, S. A., Ibáñez Elam, A. L., & Mantilla Oliveros, N. A. (2016). *La pesca ilegal marina en Colombia.* Procuraduría General de la Nación & Fundación MarViva. https://faolex.fao.org/docs/pdf/col196244.pdf

Rodríguez Ruiz, J. D. (2021). *La controversia sobre el impacto en salud de las aspersiones aéreas con glifosato en la política de erradicación de cultivos de uso ilícito: Poder, saber y política pública* [Ph.D. Thesis, Universidad Nacional de Colombia]. https://repositorio.unal.edu.co/handle/unal/80877

Romero, J. (2024, September 9). Armada colombiana detuvo a 49 personas por pescar ilegalmente en el archipiélago de San Andrés. *Infobae.* https://www.infobae.com/colombia/2024/09/09/armada-detiene-a-49-personas-por-pesca-ilegal-en-zona-protegida-del-archipielago-de-san-andres/

Sandoval Perdomo, A. E., & Arciniegas Londoño, L. (2022). Afectaciones a la seguridad y defensa de Colombia por el fenómeno de la minería ilegal: Percepción académica, escenario y posibles soluciones institucionales. *Perspectivas En Inteligencia, 13*(22), 49–61. https://doi.org/10.47961/2145194X.270

Swain, A., Bruch, C., Ide, T., Lujala, P., Matthew, R., & Weinthal, E. (2023). Environment and security in the 21st century. *Environment and Security, 1*(1–2), 3–9. https://doi.org/10.1177/27538796221149609

Torrado, S. (2021, June 9). Colombia reduce los cultivos de coca por tercer año consecutivo. *El País.* https://elpais.com/internacional/2021-06-09/colombia-reduce-los-cultivos-de-coca-por-tercer-ano-consecutivo.html

UNODC. (2020). *Colombia: Explotación de oro de aluvión. Evidencias a partir de percepción remota 2019.* https://tinyurl.com/49x5dxa8

UNODC. (2021). *Colombia: Explotación de oro de aluvión. Evidencias a partir de percepción remota 2020.* https://tinyurl.com/y5fn7fsx

UNODC. (2022). *Colombia: Explotación de oro de aluvión. Evidencias a partir de percepción remota 2021.* https://tinyurl.com/mt7hpm2r

UNODC. (2023). *Colombia: Explotación de oro de aluvión 2022. Resumen Ejecutivo.* https://tinyurl.com/4wmtbc9h

UNODC. (2024). *Crop Monitoring and Colombia.* United Nations : Office on Drugs and Crime. https://www.unodc.org/unodc/en/crop-monitoring/?tag=Colombia

Van Uhm, D. (2023). *Organized Environmental Crime: Black Markets in Gold, Wildlife, and Timber* (1st ed.). Bloomsbury Publishing Inc. https:doi.org/https://doi.org/10.5040/9798216172185

Villamil-Castillo, H. A. (2020). El pos-acuerdo, "detonante ambiental en Colombia." *Revista Academia & Derecho, 11*(21), 221–248.

Wirtz, J. (2019). A new agenda for security and strategy? In J. Baylis, J. Wirtz, & C. S. Gray (Eds.), *Strategy in the Contemporary World* (6th ed., pp. 366–384). Oxford University Press.

Luisa Fernanda Villalba García Ph.D. (c) in Strategic Studies, Security and Defense and MA in Strategy and Geopolitics, Escuela Superior de Guerra "General Rafael Reyes Prieto," Colombia. BA in International Relations and Political Studies, Universidad Militar Nueva Granada, Colombia. Professor and researcher at the Escuela Militar de Cadetes General José María Córdova, Colombia. https://orcid.org/0000-0003-3169-9312—Contact: luisa.villalba@esmic.edu.co

Andrés Eduardo Fernández-Osorio Ph.D. in Law and Political Science, University of Barcelona, Spain. MA in Economics, State and Society: Politics and Security, University College London, United Kingdom. MA in International Relations, Higher School of Economics, Russian Federation. MA in Defense Studies, King's College London, United Kingdom. BA in International Relations and Political Studies, Universidad Militar Nueva Granada, Colombia. BSc in Military Sciences, Escuela Militar de Cadetes General José María Córdova, Colombia. Adjunct Professor and researcher at the School of Humanities, Universidad Gabriela Mistral, Santiago, Chile. https://orcid.org/0000-0003-0643-0258—Contact: andres.fernandez@academico.ugm.cl

Elmers Freddy Velandia-Pardo Ph.D. in Law, Education, and Development, University of Malta, Malta. MBA, University of Valencia, Spain. MA in Human Rights and International Law of Armed Conflict, Escuela Superior de Guerra "General Rafael Reyes Prieto," Colombia. BSc in Military Sciences, Escuela Militar de Cadetes General José María Córdova, Colombia. BA in International Relations and Political Studies, and BA in Business Administration, Universidad Militar Nueva Granada, Colombia. Researcher and professor at the Universidad Militar Nueva Granada, Colombia. https://orcid.org/0000-0003-1217-9178—elmers.velandia@unimilitar.edu.co

Civil Society and the Military: Tensions and Alliances in the Construction of Security and Defense Policies

Andres Eduardo Fernandez-Osorio, Luisa Fernanda Villalba-Garcia, and Christian Acevedo-Navas

Abstract This chapter examines the complex relationship between civil society and military institutions in shaping security and defense policies in Latin America. Defining key concepts (civil society, military, and security types) and critically reviewing civil-military relations (CMR) theories alongside Security Sector Governance (SSG), it argues interaction models are more relevant than strict separation for the region. It analyzes Latin America's historical evolution, marked by militarism and transitions, identifying recurring tensions like democratic oversight versus military autonomy and internal security militarization. The analysis assesses the impact of these interactions on policy legitimacy and effectiveness, addressing emerging challenges from new threats and technologies. It concludes by emphasizing the need to strengthen democratic civilian oversight and capabilities within the security and defense sector.

Keywords Civil society · Civil-military relations · Defense and security policies · Governance · Latin America

A. E. Fernandez-Osorio (✉)
Universidad Gabriela Mistral, Santiago, Chile
e-mail: andres.fernandez@academico.ugm.cl

L. F. Villalba-Garcia
Escuela Militar de Cadetes "General José María Córdova", Bogotá, D.C, Colombia
e-mail: luisa.villalba@esmic.edu.co

C. Acevedo-Navas
Escuela Superior de Guerra "General Rafael Reyes Prieto", Bogotá, D.C, Colombia
e-mail: christian.acevedo@esdeg.edu.co

L. A. Garcia-Segura (ed.), *Collaborative Insights on Technology and Legal Frameworks*, Advanced Sciences and Technologies for Security Applications,
https://doi.org/10.1007/978-3-032-27714-5_12

1 Introduction

The study of civil-military relations (CMR) has acquired increasing relevance within the framework of contemporary democracies, particularly in regions such as Latin America, where political history has been indelibly marked by the often tense interaction between civilian power and the military establishment (Romero Herrera 2007). The interaction between civil society and military institutions is fundamental to the formulation, legitimacy, effectiveness, and long-term sustainability of security and defense policies (Sung et al. 2022). These policies, which seek to protect the State and its citizens from increasingly complex threats, are constructed in an arena where diverse interests, perceptions, and capabilities converge.

The relationship between civilian and military actors is inherently complex, characterized by a constant dialectic between tensions and alliances. On the one hand, significant frictions arise around crucial issues such as effective democratic control over the armed forces, the necessary transparency in defense management, unrestricted respect for human rights in all operations, and debates about the appropriate use of force and the risks associated with the militarization of traditionally civilian functions (Amoroso Botelho et al. 2023). On the other hand, there are also spaces and mechanisms for constructive collaboration, where sectors of civil society and the military cooperate in peacekeeping missions, humanitarian assistance, and natural disaster response or participate jointly in strategic debates and trust-building processes (Holshek and Coning 2024).

This chapter aims to analyze this dual dynamic in-depth, exploring its conceptual foundations, the theoretical frameworks that attempt to explain it, its historical evolution with an emphasis on the Latin American context, its contemporary manifestations through specific tensions and alliances, and its ultimate consequences for the governance of the security and defense sector. The key concepts of civil society, military and security and defense policies will be examined; the main theories on CMR and Security Sector Governance (SSG) will be reviewed; the historical trajectory of these interactions in Latin America will be traced; common sources of conflict and cooperation will be identified; illustrative case studies will be analyzed; the impact of these dynamics on the quality of the resulting policies will be assessed; and finally, the emerging challenges and trends that shape the future of this crucial relationship for stability and democracy in the region will be explored.

2 Definition of Key Concepts

To adequately address the complex interaction between civil society and the military in the construction of security and defense policies, it is essential to provide a sound conceptual foundation. This entails defining key terms not only formally or etymologically but also operationally, considering their historical evolution, multiple facets, and specific application in the contemporary context, with particular attention to the realities of Latin America. The way these concepts are defined is not a mere academic exercise as it reflects and, in turn, shapes power relations and perceptions about the legitimate role of each actor in the public sphere.

2.1 *Civil Society*

The concept of civil society encompasses a vast and heterogeneous range of non-state actors operating in the public sphere. Adopting the definition proposed by the Geneva Centre for Security Sector Governance (DCAF), civil society is understood as "the diverse range of groups that people form around a shared interest or a vision of the public interest" (DCAF 2015a, p. 2). These groups range from non-governmental organizations (NGOs), think tanks, and independent foundations to professional associations, trade unions, community groups, religious organizations, and social movements (DCAF 2015a).

Three main characteristics distinguish civil society in a democratic context. First, acting in the public interest as civil society articulates and expresses opinions on social and political relevance matters, seeking to influence the public agenda to improve social outcomes. Unlike political parties, their objective is not to be part of the government but to influence it and society (DCAF 2015a). Second, free organization, as participation and membership are voluntary and independent of direct government control, although they may receive state funding or interact closely with the State (DCAF 2015a). It is organized formally (associations, foundations) or informally (campaigns, networks). Third, non-profit objectives, as although they require resources to operate, their fundamental reason for being is not the generation of economic benefits (DCAF 2015a).

In the specific area of SSG, that refers to how power and authority are exercised in the provision, management, and oversight of State and human security within a country (DCAF 2015b), civil society has established itself as an indispensable actor (Arugay and Baquisal 2024). Its role goes beyond simple observation, functioning as a crucial channel for communicating the population's diverse perceptions and security needs to the State. It performs vital functions such as raising public awareness on security and defense issues; conducting independent research, analysis, and monitoring; providing complementary services (such as victim support or community policing initiatives); lobbying to influence policies; and, fundamentally, exercising public oversight, over the performance of the security sector, ensuring its adherence to the rule of law and respect for human rights (DCAF 2015a).

2.2 The Military

The term military is also multifaceted and must be analyzed from different dimensions. First, as an Institution. The Armed Forces constitute an essential institution of the State, organized to guarantee the defense of sovereignty, independence, the integrity of the national territory, and, in many cases, the constitutional order (Moncayo Gallegos 2018). They operate as a complex bureaucratic organization, are the principal armed instrument of the State's defense policy, and their primary function is the organized management and potential use of force in the name of the State (Jaimes Pinillos 2018; Soprano 2010).

Second, as a Class/Profession. Military sociology has extensively studied the characteristics of the military profession. Samuel Huntington defines it by its expertise (specialized knowledge), corporatism (sense of unity and collective identity), and responsibility (service to society and subordination to political authority) (Huntington 1957). For his part, Morris Janowitz qualifies the idea of political neutrality, viewing the military as an effective pressure group and analyzing the tension between the ethos of the "heroic warrior" and more administrative or managerial functions (Janowitz 1974). The military profession is often associated with a strong vocation for service and values such as honor, patriotism, and fulfilling duty (Fernandez-Osorio 2021). Charles Moskos distinguishes between an "institutional" model (based on values and service) and an "occupational" one (based on the labor market), a tension present in modern armed forces (Moskos 1986; Segal 1986).

Third, as a Political Actor. It is crucial to recognize that, beyond its institutional and professional role, the military establishment has historically been and continues to be a fundamental political actor, especially in Latin America (Droz-Vincent 2014; Remmer 2025). Its monopoly or quasi-monopoly of organized force gives it a unique capacity to influence and pressure the political system, even in democratic regimes (Agüero 2001; Jenne and Martínez 2022). Latin American history is replete with examples of direct (coups d'état) and indirect (tutelage, policy influence) military intervention (Tirado 2021; Victoriano Serrano 2010). This inherent political dimension, which clashes with the normative ideal of military subordination to democratically elected civilian power, is a constant source of tension and a key factor in understanding the particularities of CMRs in the region. Ignoring this political facet would lead to an incomplete and potentially naive analysis of civil-military dynamics.

2.3 Security and Defense Policies

This concept has also undergone a notable evolution. First, as a traditional vision (National Defense). Historically, defense policies focused on protecting the nation-state from external military threats to preserving sovereignty and territorial integrity, and defense management was concerned with organizing resources (human, financial, material) to accomplish this mission (Bucur-Marcu et al. 2013; Cano Cuevas 2024). Second, as a contemporary expansion of the concept. The end of the Cold War and the emergence of new threats and actors have led to a significant expansion of the security concept (Bartolomé 2018).

For instance, the definition of national security has expanded to include the protection of the State and the population from internal threats (such as organized crime or terrorism) in addition to external ones (Cabrera-Ortiz 2021; Monroy Hernández et al. 2013); and human security has emerged as an alternative or complementary paradigm, focusing attention on the security of the individual ("freedom from fear" and "freedom from want"), encompassing dimensions such as physical, economic, food, health, environmental, and personal security. Community and politics help protect human rights (Acevedo-Navas et al. 2022). This approach questions the absolute primacy of state security when it conflicts with the security of individuals (Orozco 2005).

Other definitions have risen as well. Citizen security focuses on the protection of citizens from common crime and violence in public spaces and is a central concern in Latin America (Muggah 2017); societal security refers to the "maintenance and preservation of the language, culture, religion and ethnic identity of a given community" (Buzan et al. 1998, as cited in Orozco 2005, p. 169); and comprehensive security that seeks to encompass all dimensions of security (external, internal, human, and citizen) within a unified framework, as mentioned in contexts such as Ecuador (Moncayo Gallegos 2018).

The polysemy and evolution of these key concepts—civil society, military, and security and defense policies—are not merely semantic. They reflect profound transformations like the modern State, the changing perception of threats (from interstate to transnational and human), and, crucially, the power relations between various State and non-state actors (Bartolomé 2018). The conceptual tension between state-centered security (national security/defense) and individual-centered security (human security) is a field of ideological and political contestation. The prioritization of one concept over the other has direct implications for the allocation of resources, the definition of military missions, the type of public policies implemented, and, ultimately, the model of civil-military relations that is fostered or considered legitimate. Therefore, the definition and use of these concepts are acts with significant strategic consequences.

3 Theoretical Frameworks in Civil-Military Relations

Numerous theoretical analyses have examined the complex relationship between civil society and military institutions in security and defense policymaking. Understanding these frameworks is essential to interpreting the dynamics of tension and collaboration. A critical literature review reveals approaches from classical CMR theories to more recent SSG paradigms.

3.1 Classical Theories of Civil-Military Relations

These theories, which emerged primarily in the Western context of the postwar and Cold War, sought to explain how to maintain civilian control over professional and powerful military institutions. Samuel P. Huntington, in his influential work The Soldier and the State, argued that the best way to ensure civilian control is by maximizing military professionalization (objective control) (Huntington 1957; Travis 2017). A highly professional officer corps, focused on its specific expertise (the management of violence), with a strong corporatism and conscious of its responsibility to the State, would tend to stay out of partisan politics, accepting subordination to legitimate civilian authority.

This objective control is based on creating distinct and separate spheres of competence: civilians define the political objectives of war and security, while the military autonomously determines the operational means to achieve them. However, this theory has been criticized for its limited applicability in contexts where professionalization is low, military politicization is high, or threats are predominantly internal (Skaar and Malca 2014). Furthermore, it has been objected that it normatively restricts citizen and civil society participation in defense matters, considering them the exclusive domain of military experts and political elites.

Morris Janowitz, with his constabulary force concept, offered a different perspective, questioning the inherent political neutrality of the military and viewing it more as an interest group within the state apparatus (Nielsen and Liebert 2021; Travis 2020). He proposed that, with the evolution of warfare and technology, the military role shifted from a "heroic warrior" model to one closer to a constabulary force involved in a broader range of tasks, including maintaining internal order and crisis management. He argued that the armed forces were becoming more permeable to societal influences and that a sure convergence or closeness with civil society was inevitable and, in certain aspects, desirable to maintain legitimacy. Although he also advocated for the functional separation between the civilian and military sectors in politics, his approach recognizes greater interpenetration.

Other classical theorists, such as Samuel Finer, developed typologies of the levels of military intervention in politics, from influence to the replacement of civilian rule (Finer 1969; Sandhu 2016). Alfred Stepan analyzed military prerogatives (areas of autonomy or tutelary power) that often persist even after transitions to democracy, especially in Latin America, and the importance of legal reforms to limit them (Stepan 2001, 2020). Charles Moskos introduced the dichotomy between the "institutional" (values- and service-based) and the "occupational" (market-based) models to understand the motivations and culture within the armed forces (Moskos 1986; Segal 1986; Taylor et al. 2015).

3.2 Contemporary Civil-Military Relations Theories

More recent approaches have sought to refine or challenge classical models, incorporating new variables and contexts. Peter Feaver and his Agency Theory apply principal-agent economic theory to CMR. Civilian leaders are the "principals" who delegate security tasks to the military, the "agents" (Feaver 1998, 2003). The central problem is ensuring the military (agents) comply with civilian directives (a behaviour Feaver terms "working") rather than following their preferences or evading oversight (which he terms "shirking"). Civilian control is exercised through monitoring and the application of sanctions. This theory has been criticized for its potentially conflictual and transactional view of the relationship and for not adequately capturing elements such as trust, loyalty, and shared values (Krupski 2023). As an alternative, the Centripetal Guardianship Model has been proposed, based on stewardship theory, which suggests that the military acts as "guardians" of constitutional values and national defense, motivated by shared goals with civilian leaders and by mutual trust rather than by coercion or strict control (Day and Khisa 2022; Uluçakar and Çağlar 2017).

Rebecca Schiff and her Concordance Theory propose an alternative model especially relevant to non-Western contexts (Schiff 1995, 2008). She argues that the stability of CMR depends not so much on strict separation (Huntington) as on "concordance" or agreement between three key actors: the military, the political elite, and the citizenry. When these three actors share fundamental values about the role of the armed forces, see each other as partners in providing security, and maintain dialogue and mutual accommodation, military intervention in domestic politics is less likely (Schiff 2008). This theory is particularly relevant to the topic of this chapter, as it gives a central and explicit role to citizens (and, by extension, organized civil society) in shaping CMR, something that classical theories often overlook.

Finally, some frameworks that center on the SSG and Security Sector Reform (SSR), primarily driven by international organizations and the development community since the 1990s, represent a more holistic and integrated approach (Born et al. 2000): a "Good SSG" involves applying the principles of good governance (accountability, transparency, rule of law, participation, responsiveness, effectiveness, efficiency) to the security sector (Born et al. 2000). SSR is the political and technical process that promotes good SSG and improves State and human security. This approach considers a broad spectrum of actors, including the military, police, intelligence, criminal justice, management institutions (ministries), and oversight actors (parliaments, civil society).

Unlike many classical CMR theories, SSG/SSR frameworks explicitly assign a pivotal role to civil society. It is considered a key actor for democratic oversight, promoting accountability, articulating citizen demands on security, contributing to informed debate, and promoting accountability on policies, and even as an alternative security provider in specific contexts. The roles identified for civil society include being an agent of accountability, creating spaces for new discourses on security and development, and acting as an alternative provider of people-oriented security (Arugay and Baquisal 2024).

An analysis of these theoretical frameworks reveals fundamental tension. Classical theories, often developed in established Western democracies, seek mechanisms to control or isolate a military establishment perceived as potentially threatening to civil order. However, Latin America's historical and contemporary reality shows a profound interweaving of the military into the political and social sphere, with democracies often fragile or consolidating and persistent authoritarian legacies. This suggests that models such as Huntington's, based on strict separation, may have limited applicability. In contrast, theories such as the Schiff Concordance, which emphasize interaction and agreement among multiple actors (including citizens) (Schiff 2008) or SSG/SSR approaches, which promote shared governance and the active participation of civil society, could offer more appropriate analytical tools for understanding and, potentially, improving CMR in the region.

The emergence of the SSG and SSR frameworks represents, in fact, a paradigm shift. By explicitly integrating the dimensions of security, development, and democratic governance and assigning a central normative role to civil society in oversight and accountability, these approaches go significantly beyond what classical CMR theories contemplate. They offer an idea of how CMR should function in a modern democracy. However, their practical implementation in Latin America and other regions faces considerable obstacles, including a lack of political will on the part of elites, resistance from security sector actors themselves, state capacity constraints, and, crucially, the heterogeneity, capacity, and degree of pluralism of civil society itself to assume the roles assigned to it effectively.

4 Historical Evolution of Civil-Military Interactions in Latin America

Understanding the current tension dynamics and alliance between civil society and the military in Latin America requires a historical perspective. Their interactions have been complex, marked by cycles of military intervention, fragile democratic transitions, and constant renegotiation of roles and powers.

4.1 The Age of Militarism and the National Security Doctrine (Until the Late 1970s)

For much of the twentieth century, Latin American armed forces played a leading political role, frequently intervening in politics to overthrow civilian governments or supervise the political system (Velázquez 2008). This phenomenon, often referred to as militarism (Telleria 2024), intensified during the Cold War with the widespread adoption of the National Security Doctrine (NSD) (Skaar and Malca 2014). Primarily influenced by the United States and French counterinsurgency schools in their fight against communism, the NSD redefined the primary security threat as internal (subversion) and, particularly in the Southern Cone, justified the seizure of power by the armed forces to "protect" the nation from this internal enemy (Álvarez Veloso et al. 2012). This period was characterized by military dictatorships in many countries of Southern Cone and Central America, with human rights violations and the suppression of organized civil society (Cardenas and Root 2022). The civil-military relationship was one of apparent subordination of the civilian to the military, with little or no space for citizen participation or oversight.

4.2 Transitions to Democracy (1980s–1990s)

Beginning in the late 1970s and throughout the 1980s, most Latin American military regimes entered crisis and gave way to democratic transition processes, part of what Huntington (1991) termed the third wave of democratization, as applied to the regional case by Velázquez (2008); however, the nature of these transitions varied significantly. In some cases, the transition resulted from a military defeat (Argentina) or regime collapse, allowing for a clearer break with the authoritarian past. In many others, however, the transition was negotiated or pacted between civilian and military elites (Germani and Silvert 1961; Rouquié 1987).

These modes of transition had a lasting impact on post-authoritarian CMR as packed transitions often allowed the armed forces to retain important amounts of power and autonomy (military prerogatives), such as special judicial privileges, control over their budgets, influence on key appointments, or even tutelary roles enshrined in constitutions (Avilés 2009; Jaskoski 2012; McSherry 1995). Establishing effective democratic civilian control over military institutions that perceived themselves as saviors of the homeland and had not been defeated became one of the most significant challenges for the new democracies. Civil society, which in many cases had played an important role in the resistance against dictatorships, now faced the challenge of consolidating spaces for participation in an often tutelary democratic context (Silva 2002).

4.3 Democratic Consolidation, New Roles, and the Persistence of Old Problems (Post-1990s/2000s)

With the end of the Cold War and the progress of democratization, Latin American armed forces were forced to redefine their missions (Pion-Berlin 2016) as the communist threat disappeared, and interstate conflicts became less likely (Jenne and Martínez 2022; Ross 2004). In this new scenario, new perceived threats emerged, such as drug trafficking, transnational organized crime, terrorism, and, more recently, natural disasters or the protection of strategic resources (Acevedo-Navas 2023). This led to an increasing involvement of the armed forces in internal security tasks and non-traditional roles, such as supporting the police, counter-drug enforcement, disaster relief, and even participation in development projects (Galindo García and Aristizábal Murillo 2021).

While progress was made in formal civilian control in many countries (e.g., appointment of civilian defense ministers and legal reforms), significant deficits persisted. Military autonomy in certain areas remained considerable, and parliamentary and judicial oversight was often weak (Sotomayor Velásquez 2006). The emergence of citizen security, focused on protecting individuals from common crime, generated new debates on the relevance of military participation in traditionally police-related tasks (Luque-Juárez 2024; Quintero Cordero 2020).

4.4 Recent Trends: Backsliding and "Civilian Militarism"? (Post-2010)

Various analysts have pointed to worrying trends in the region's CMRs in the last decade. Faced with persistent public security crises, high levels of criminal violence, and, in some cases, political and institutional crises, several democratically elected governments have increasingly turned to the armed forces for law enforcement and other civilian functions (Morlino 2014). This phenomenon has been termed "new militarism" or "civilian militarism," where civilian authorities themselves, for pragmatic reasons, popularity, or institutional weakness, invite or expand the military's role in the domestic sphere (Diamint 2015; Rodriguez 2025). This has revived debates about the militarization of security, its effects on human rights, and the quality of democracy.

Therefore, the history of CMR in Latin America cannot be interpreted as a linear progression from authoritarianism to democracy with complete civilian control (Velázquez 2008). Instead, it resembles a cyclical and complex process, where the progress achieved coexists with the persistence of authoritarian legacies (Dammert 2019) and the emergence of new forms of military influence or intervention, often under the auspices of democratic legality (Robledo Hoecker 2022). The consolidation of democratic civilian control remains an unfinished and, in some respects, fragile and potentially reversible task in the region (Sotomayor Velásquez 2006).

A key factor underlying this complexity is the cultural and institutional legacy of the NSD. However, the NSD, as such, was ideologically superseded after the Cold War; the "internal enemy" mentality and the justification for military intervention in domestic affairs to preserve "order" appear to have endured (Álvarez Veloso et al. 2012). The logic previously applied to "communist subversion" is now being repurposed to frame the fight against organized crime, drug trafficking, or, in some cases, even social protest, as existential threats requiring a militarized response. This cultural persistence facilitates the securitization of social and political problems and paves the way for expanding military roles to the detriment of civilian and democratic solutions.

5 Common Sources and Manifestations of Tension

The relationship between civil society and military institutions in Latin America is marked by recurring tensions stemming from historical, institutional, and cultural sources. These frictions manifest themselves in various security and defense policymaking and implementation areas.

5.1 Democratic Control Versus Military Autonomy

One of the most fundamental tensions lies in defining and exercising democratic civilian control over the armed forces (Amoroso Botelho et al. 2023). While the democratic principle demands the military's subordination to legitimately constituted political power, the armed forces often seek to preserve significant degrees of institutional autonomy.

This tension manifests itself in different ways. First, resistance to oversight makes it difficult for civilian branches (executive, legislative, judicial) to supervise military decisions and operations effectively. Second, defense of prerogatives as the military may consider certain areas within its exclusive jurisdiction (doctrine, promotions, military justice, internal management, operating budget), resisting civilian interference. Third, civilian militarism, as in some cases, civilian elites, out of weakness or political calculation, voluntarily cede space and power to the military, reversing the logic of control.

5.2 Transparency and Accountability

Opacity is a constant source of tension. The defense sector, often protected by national security arguments, tends to be less transparent than other areas of the State (Gorbanova and Wawro 2011; Villalba-Garcia 2022). This translates into budgetary opacity, lack of information on operations, and impunity.

On the one hand, there is a lack of detail and public scrutiny over how the defense budget is spent, especially in material procurement. This can hide inefficiency or corruption (Stone 2019). Tensions surrounding budgetary transparency are a critical issue, as budget control is a key tool of civilian power. Opacity limits the ability of civil society and parliaments to exercise effective oversight. On the other hand, the lack of information on operations creates difficulties in accessing accurate information about military operations, especially those involving the use of force or impacting the civilian population. Additionally, significant obstacles to investigating and punishing members of the armed forces responsible for human rights violations or other crimes. The persistence of military jurisdiction or the lack of collaboration with the civilian justice system contributes to a culture of impunity that erodes public trust. The lack of adequate external oversight mechanisms (parliamentary, judicial, social) exacerbates this problem (De Vicente Encarnación 2024).

5.3 *Militarization of Public Security and Civil Society-Military Mistrust*

The growing trend toward using the armed forces to combat organized crime, patrol streets, or perform other internal security tasks is one of the most significant sources of contemporary tension. Criticisms focus on ineffectiveness as doubts about whether militarization reduces violence and crime in the long term; the impact on Human Rights as sometimes a correlation between military presence in internal security and increased abuses; institutional weakening as a disincentive to reform and strengthen civilian police forces; and democratic erosion as normalization of military roles in the civilian sphere, weakening democratic oversight and the rule of law.

There is often a gap in understanding and trust between sectors of civil society and military institutions. This is fueled by historical legacies in the form of memory of dictatorial repression and abuses that generate mistrust in broad sectors of society (Cremades 2021); mutual stigmatization as some civilian actors tend to view all military personnel as potential human rights violators, while some military personnel may perceive human rights NGOs or critics as enemies or obstacles; and lack of dialogue and the exchange of perspectives that allow prejudices to be overcome and mutual understanding to be built.

It is essential to understand that these tensions do not operate in isolation but are deeply interconnected and often reinforce each other, creating vicious cycles. For example, the decision to militarize public security, often justified by police weakness and citizen pressure in the face of crime, can lead to an increase in human rights violations due to military training and doctrine. The lack of effective accountability mechanisms and the resulting impunity allow these abuses to persist, eroding citizens' trust in the security forces and the State itself. This loss of legitimacy hinders the citizen cooperation necessary for crime prevention and can lead to demands for stricter decisions, perpetuating the cycle of militarization and tension. Addressing these frictions, therefore, requires a systemic approach that recognizes these interdependencies and seeks comprehensive solutions rather than isolated interventions.

6 Areas and Mechanisms of Partnership or Cooperation

Despite inherent tensions, the relationship between civil society and military institutions is not exclusively conflictual. Important areas and mechanisms exist through which constructive interactions, partnerships, and cooperation are developed in pursuit of common objectives related to security, peace, and social well-being.

6.1 Peacekeeping Operations (PKO) and Humanitarian Assistance and Disaster Response (Civil Defense)

United Nations PKOs constitute a key arena for Civil-Military Cooperation (CIMIC). The armed forces of several Latin American countries actively participate in these missions. UN doctrine defines CIMIC as a system of interaction that includes negotiation, mutual support, joint planning, and information exchange between the military components of the mission and a wide range of civilian actors: humanitarian organizations, development agencies, local authorities, and the civilian population (Mockaitis 2004). CIMIC's primary functions include liaison and information management to facilitate communication and mutual understanding between military and civilians; mission support so the military component can provide logistical or security support to civilian actors to facilitate their tasks; and community support to guarantee the implementation of quick-impact projects (QIPs) to address urgent needs of the local population and build trust.

Given Latin America's vulnerability to natural disasters (earthquakes, floods, hurricanes, etc.), the armed forces' participation in civil defense and disaster risk management tasks is a significant area of civil-military cooperation (Inter-American Defense Board 2024b). The Armed Forces usually intervene in the preparation, response (search and rescue, evacuation, initial medical care, security, logistical transport), and, sometimes, reconstruction phases (Jiménez Espinosa and Espinosa Gallardo 2020).

This work requires close coordination with national civilian organizations, such as INDECI in Peru or Civil Protection (Chavez Cresta 2021; Novak-Talavera 2024) and international organizations (OCHA, Red Cross). Some countries have developed specialized military units for these tasks, such as the Military Emergency Unit (UME) in Spain (Expósito 2024), or have developed specialized plans and doctrines to foster better integration with civilian institutions, as in Argentina (Ugarte 2024), Brazil (Janot et al. 2024), Chile (Baeza-Freer and Escudero-Illanes 2024), Colombia (Guzmán Mesa 2016), Dominican Republic (Bobea 2024), El Salvador (Molina-Gabriel 2024), Guatemala (Bueno and Martínez 2024), Mexico (Barrachina-Lisón and Benítez-Manaut 2024) and Uruguay (Besio-Echeverría 2024).

6.2 *Confidence-Building Measures and Strategic Debates and Public Policies*

Confidence and Security Building Measures (CSBMs) are diplomatic and military tools designed to increase transparency, reduce uncertainty, and prevent state misunderstandings, contributing to regional stability (Inter-American Defense Board 2024a). Although their primary focus is interstate, they can also have a civil-military component by promoting dialogue and openness. The Inter-American Defense Board, for example, organizes workshops and facilitates the exchange of information on defense policies (such as White Papers) to foster mutual trust among member countries and, potentially, between their defense sectors and civil societies (Inter-American Defense Board 2024a).

Defense White Papers are official documents that outline a country's defense policy, threat perceptions, strategic objectives, and the structure of its armed forces (Pacheco Gaitán 2016). They are presented as instruments for increasing transparency to communicating defense intentions and capabilities to society and the international community; promoting public debate to encourage the participation of civil society, academia, and the political class in security and defense matters; building consensus and legitimacy-seeking social and political support for defense policies and fostering a defense culture (Calduch 2020); and building mutual trust to serve as a confidence-building measure by clarifying strategic positions.

Some Latin American countries, such as Paraguay (Ministerio de Defensa Nacional 2013), Ecuador (Resdal 2018), and the Dominican Republic (Presidencia de la República Dominicana 2023), have published White Papers, often stating that they are the product of a participatory process that includes civil society and academia. Beyond formal mechanisms, there may be specific initiatives for dialogue and rapprochement between military institutions and civil society organizations (universities, NGOs, community groups) to overcome historical mistrust, exchange perspectives, and find convergence points. Although often informal or localized, these spaces can be important for building bridges and improving mutual understanding.

It is important to analyze these areas of cooperation critically. They are often presented as inherently positive and depoliticized, such as disaster relief or participation in PKOs. However, these activities can be instrumentalized by governments to improve the public image of the armed forces or justify high budgets and by the military itself to maintain its relevance and capabilities in the absence of conventional conflict. While positive cooperation may have tangible benefits, it can also divert attention and resources from the need to strengthen the capacities of civilian institutions (such as civil protection systems or police forces) (Jenne and Martínez 2022). A complete analysis must go beyond the description of the cooperation and assess its underlying political and strategic implications.

Similarly, Defense White Papers, although promoted as tools for transparency and dialogue, require scrutiny. Their real democratizing value lies in their publication and the authenticity and depth of the participatory process that underpins them. It is necessary to evaluate on a case-by-case basis whether they reflect a genuine consensus built with significant participation from civil society and academia or are primarily strategic communication exercises by the military establishment or the government in power, with limited real external influence.

7 Impact of Civil-Military Interactions

The way civil society and military institutions interact has direct and profound consequences for the resulting security and defense policies. It is crucial to assess this impact not only in terms of operational outcomes but also considering democratic legitimacy, the actual effectiveness of security provision, and the long-term sustainability of these policies and the relationships that underpin them.

7.1 Impact on Legitimacy

The legitimacy of security and defense policies and the institutions that implement them (military, police, government) are strongly influenced by the nature of the CMR. Conflictual interactions, especially those marked by abuses of power, human rights violations, and lack of accountability on the part of military or police forces, seriously erode citizen trust. The militarization of public security, although sometimes initially supported by fear of crime, tends to generate a loss of trust in the medium and long term, particularly in the police, perceived as ineffective or overwhelmed, and eventually also in the military if they are implicated in abuses. Conversely, cooperation perceived as positive (e.g., disaster response) and transparent and participatory processes (e.g., drafting White Papers with genuine consultation) can strengthen trust and legitimacy.

Policies formulated and implemented with little transparency, without effective civilian oversight, or with violations of the rule of law and human rights lack democratic legitimacy, regardless of their operational results. The effective subordination of the military to elected civilian power and respect for democratic procedures are fundamental to the legitimacy of the security sector. Expanding military roles in the civilian sphere, especially if normalized, can be perceived as an erosion of democracy. Evidence from Latin America suggests sometimes a negative correlation between the militarization of internal security functions and democratic legitimacy (Sung et al. 2022). Although surveys may reflect greater initial trust in the armed forces compared to the police, sustained involvement in policing, often accompanied by reports of abuse and a lack of precise results in reducing violence, tends to undermine that trust and call into question the legitimacy of the model.

7.2 Impact on Effectiveness

Evaluating the effectiveness of security and defense policies requires going beyond purely military or operational metrics and raises some questions. For instance, do the policies succeed in reducing violence, protecting territory, and dismantling criminal organizations? The evidence on the effectiveness of the militarization of public security is sometimes mixed and often negative in the long term. While there may be specific tactical successes, they do not always translate into sustainable improvements in citizen or human security. Similarly, is allocating resources between civilian and military actors the most efficient way to achieve security objectives? (Perito 2019). Duplication of functions, lack of coordination, or the assignment of tasks to non-specialized institutions (e.g., military personnel performing police work) can generate inefficiencies. Opacity in military spending also makes it challenging to assess efficiency.

A policy may be operationally effective in eliminating a perceived threat but profoundly ineffective if it does so at the cost of massive human rights violations, weakening the rule of law and generating greater insecurity for the civilian population. The case of the "false positives" in Colombia is a tragic example: they were presented as military successes (enemy casualties), but they constituted crimes against humanity that undermined the legitimacy of the armed forces and the State (Barbosa Castillo and Ciro Gómez 2020; Cepeda-Jaramillo et al. 2021). Therefore, effectiveness must be measured multidimensionally, integrating respect for democratic principles and human rights as fundamental criteria, not as externalities.

7.3 Impact on Sustainability

Sustainability refers to the capacity of policies and institutional structures to endure and function adequately over the long term and raises questions. For example, do civil-military interactions and resulting policies strengthen or weaken the overall capacity of the State and its democratic institutions? Overreliance on the armed forces for civilian tasks can stunt capacity development in other agencies (police, civil protection, etc.), creating unsustainable dependency. Similarly, do policies address the root causes of insecurity and conflict, or do they simply manage the symptoms, perpetuating cycles of violence and militarization? (United Nations 2023). Lack of investment in prevention, social development, and judicial reform while prioritizing military response can be unsustainable in the long term.

Finally, does the CMR framework allow for flexible adaptation to new challenges and threats, or does it generate institutional rigidities? In general, dynamics of excessive tension and lack of democratic oversight tend to generate less legitimate policies, potentially ineffective in terms of human security, and unsustainable by eroding democratic foundations. On the other hand, civil-military cooperation based on transparency, mutual respect, and subordination to civilian power can generate more legitimate, effective (in the broadest sense), and sustainable policies. However, as noted above, even cooperation must be critically analyzed to ensure that it does not hide power imbalances or serve to avoid necessary reforms in the civilian sector. The final impact assessment requires a nuanced analysis that considers the specificities of each context and weighs the different dimensions of legitimacy, effectiveness, and sustainability.

8 Exploring Current Challenges and Emerging Trends

The security and defense landscape, and with it, civil-military relations, is constantly evolving. A series of emerging challenges and trends are reshaping this field, raising new questions about the roles, capabilities, and control of civilian and military actors in the twenty-first century, especially in Latin America.

8.1 New Threats and Their Impact

The nature of security threats has diversified beyond traditional interstate conflicts. The persistence and growing sophistication of networks dedicated to drug trafficking, human trafficking, arms trafficking, illegal mining, and other illegal activities by Transnational Organized Crime (TOC) represents a significant challenge to state sovereignty and citizen security in Latin America (Ruiz Campillo and Montenegro Escánez 2019). The ability of TOC to corrupt institutions and exert territorial control in some areas continues to pressure governments to employ the armed forces, keeping tensions about militarization alive (Aguirre and Frederic 2024).

The effects of climate change, more frequent and intense extreme weather events, rising sea levels, droughts, generating natural disasters, competition for scarce resources (water, fertile land), forced population displacement, and potential social and political instability create new roles and demands for the armed forces in emergency response, humanitarian assistance, and adaptation tasks. However, they can also exacerbate tensions if the response is inequitable or if resource management is militarized. In addition, cyberattacks against critical infrastructure (energy, finance, communications), massive data theft, online disinformation and propaganda, and cyberespionage are growing threats to state stability, the economy, and public trust. This calls for urgently developing civilian and military cyber defense capabilities and increased international cooperation (Gobierno de Chile 2019). This domain's technical and often opaque nature poses challenges for democratic oversight.

8.2 Impact of Emerging Technologies

Rapid technological evolution is transforming the nature of security and warfare, with profound implications for CMR. Artificial Intelligence (AI) and its application in the military field is a vast and growing development of Lethal Autonomous Weapons Systems (LAWS), improved intelligence analysis, logistics optimization, cognitive warfare (disinformation), surveillance, and reconnaissance (Abaimov and Martellini 2020; Longpre et al. 2022). This raises fundamental ethical dilemmas regarding meaningful human control over lethal force, accountability in the event of errors, and the risk of algorithmic bias (Micha and Farías 2021). There is fear of a new arms race and strategic destabilization.

Similarly, drones (UAVs) and their proliferation for surveillance, reconnaissance, and armed attacks is a global reality, including in Latin America (Janot and de Araujo de Assis 2025; Sánchez 2014). Their use raises serious questions about privacy and human rights, lowering the threshold for using force and the urgent need for legal and ethical frameworks for their regulation (Jorquera Escobar 2024). Finally, combining AI, big data, facial recognition, and other technologies enables surveillance and mass control levels. This has direct implications for civil liberties, privacy, and the potential for abuse of power, challenging democratic oversight.

These new threats and technologies converge to create a more complex and potentially opaque security and defense environment. For example, the technical nature of cybersecurity or AI can make it difficult to understand and oversee civilian actors (parliaments, civil society, the judiciary) who lack the necessary expertise. This creates the risk of a growing gap between those who possess technical-military knowledge and those who should exercise democratic oversight, weakening accountability. The future of civilian oversight will crucially depend on the ability of these actors to adapt, acquire knowledge, and develop new oversight mechanisms appropriate to this new technological and strategic reality.

8.3 Sociopolitical Changes, Governance, and Trends in SSG

The sociopolitical context in Latin America also presents challenges: a crisis of political representation, high citizen distrust in institutions, including political parties and sometimes the State itself, growing ideological polarization, and a more mobilized and demanding citizenry (demands for better services, security, and an end to corruption). Events like the COVID-19 pandemic have also impacted the CMR, often leading to increased military visibility and participation in control and logistics tasks (Hamilton and Bordón 2023). This sociopolitical volatility can influence the stability of the CMR, either by generating popular demands for order that favor militarized solutions or by provoking tensions if the armed forces are perceived as repressive or aligned with unpopular elites (Schmidt 2024).

At the normative and international levels, there remains an emphasis on the principles of SSG: democratic civilian oversight, accountability, transparency, respect for human rights and the rule of law, and citizen participation (DCAF 2015b). The importance of national ownership of reforms (local leadership and commitment) is increasingly recognized to ensure their sustainability. However, effective implementation remains a challenge. At the same time, potential tension arises with geopolitical approaches such as "integrated deterrence" promoted by the US (Aguirre and Frederic 2024), which prioritizes the alignment of national capabilities against global strategic competitors (China, Russia). For Latin America, this involves navigating between external geopolitical pressures, which may favor specific types of cooperation or military capabilities, and internal agendas for democratic security sector reform, which may have different priorities (e.g., police reform, transitional justice, human security) (Godnick 2023). How countries in the region manage this tension between external and internal imperatives will be crucial to the future of their CMR and security policies.

Finally, there is a growing need to incorporate strategic foresight into security and defense planning in Latin America (Marczak et al. 2016). Anticipating future scenarios, analyzing long-term trends, and adapting foresight methodologies, such as those used by the US or Australia, with adjustments to the regional context, can help decision-makers better prepare for uncertainty and make more informed decisions in the present.

9 Conclusions

The analysis of the interaction between civil society and military institutions in constructing security and defense policies reveals an intrinsically complex, dynamic, and often contradictory relationship, especially in the Latin American context. Far from being separate actors that occasionally interact, civil society and the military are intertwined in a constant dialectic of tensions and alliances that profoundly shape the governance of the security sector.

The very definition of key concepts—"civil society," "military," and "security"—is not neutral but rather reflects and configures power relations and normative visions about the role of each actor. The persistence of the "military" as an inherent political actor in Latin America, beyond its professional function, is crucial for understanding regional dynamics. Classic CMR theories, focused on control and separation, show limitations in Latin America. Approaches such as Concordance Theory and, especially, SSG/SSR frameworks, which emphasize interaction, shared governance, and the active role of civil society, appear to offer more relevant analytical tools. However, their practical implementation faces significant challenges.

The historical evolution of CMR in the region has not been linear. Democratic advances coexist with authoritarian legacies and new forms of militarization, making democratic civilian control a fragile and often reversible process. The sources of tension are not isolated but interconnected in vicious cycles that can undermine security policies' legitimacy, effectiveness, and sustainability. Areas of partnership and cooperation, including peacekeeping operations, disaster response, CSBMs, and White Papers, can generate positive results, but they can also be instrumentalized to improve institutional images, justify budgets, or avoid necessary structural reforms in the civilian sector. These areas therefore require critical analysis.

Assessing the impact of civil-military interactions must be multi-dimensional. Militarized policies may demonstrate short-term operational effectiveness, but they often generate high costs regarding democratic legitimacy, respect for human rights, and long-term institutional sustainability. Militarization frequently erodes citizen trust. New threats (TOC, climate, cyber) and emerging technologies (AI, drones, surveillance) add complexity and opacity to the security sector, challenging existing civilian oversight and accountability mechanisms. Navigating the tensions between external geopolitical pressures and internal democratic reform needs will be a key challenge for the region.

The findings of this analysis have significant implications for both theory and practice. Theoretically, they underscore the need for analytical frameworks that capture civil society's complexity, contextual contingency, and active role in CMR, overcoming overly simplistic or normative views. Practically, they warn of the risks inherent in militarizing security and the importance of strengthening civilian institutions, transparency, and democratic oversight mechanisms.

For policymakers, the challenge lies in designing and implementing security and defense policies that are effective against real and legitimate threats in the eyes of citizens while fully respecting the rule of law and human rights. This requires resisting the temptation of easy militarized solutions and instead investing in the long-term strengthening of civilian capacities (police, judicial, civil protection) and the promotion of preventive and comprehensive approaches.

Civil society's task is to maintain and strengthen its role of critical oversight, independent investigation, and promotion of democratic alternatives. This requires developing technical capabilities (including an understanding of new threats and technologies), fostering collaboration between diverse organizations, and seeking effective spaces for dialogue and political advocacy, overcoming historical mistrust and potential co-optation.

Ultimately, building legitimate, effective, and sustainable security and defense policies in the twenty-first century depends on a continued commitment to informed dialogue, radical transparency, and the mutual strengthening of democratic oversight between civil society, its political representatives, and military institutions. Only through this vigilant and constructive interaction can we hope for lasting peace and security, anchored in democratic principles. Future research should delve deeper into the specific impact of new technologies on civilian oversight, conduct more detailed comparative studies of successful (and unsuccessful) strategies for civil society engagement in security sector reform, and analyze the effectiveness of regional cooperation mechanisms in the face of transnational threats.

References

Abaimov, S., & Martellini, M. (2020). Artificial Intelligence in Autonomous Weapon Systems. In *21st Century Prometheus* (pp. 141–177). Springer. https://doi.org/10.1007/978-3-030-28285-1_8

Acevedo-Navas, C. (2023). Ejes temáticos estratégicos en seguridad y defensa en Colombia. *Revista Científica General José María Córdova, 21*(42), 303–326. https://doi.org/10.21830/19006586.1215

Acevedo-Navas, C., Ballesteros-Betancur, V., & Corcione-Nieto, M. A. (2022). Seguridad humana y seguridad multidimensional, su enfoque y utilidad para proteger los derechos humanos. *Revista Científica General José María Córdova, 20*(40). https://doi.org/10.21830/19006586.1081

Agüero, F. (2001). Institutions, Transitions, and Bargaining: Civilians and the Military in Shaping Post-Authoritarian Regimes. In D. Pion-Berlin (Ed.), *Civil-Military Relations in Latin America* (pp. 194–222). University of North Carolina Press.

Aguirre, M., & Frederic, S. (2024). *Tendencias de la Seguridad en América del Sur*. Friedrich-Ebert-Stiftung. https://library.fes.de/pdf-files/bueros/la-seguridad/21050.pdf

Álvarez Veloso, D., Garzón Vergara, J. C., Tellería Escobar, L., & Fiumara, M. P. (2012). *El papel de las Fuerzas Armadas en América Latina. Seguridad interna y democracia.* CLACSO.

Amoroso Botelho, J. C., Arciniegas Carreño, A., & González Tule, L. A. (2023). Más allá de la seguridad pública: la militarización de la política democrática en América Latina. *Canadian Journal of Latin American and Caribbean Studies, 48*(2), 279–300. https://doi.org/10.1080/08263663.2023.2190654

Arugay, A. A., & Baquisal, J. K. A. (2024). Introduction. In *Accountability, Discourse, and Service Provision: Civil Society's Roles in Security Sector Governance and Reform (SSG/R) and Sustainable Development Goal-16 (SDG-16)* (pp. 1–7). Ubiquity Press. https://doi.org/10.5334/bcy.a

Avilés, W. (2009). Despite Insurgency: Reducing Military Prerogatives in Colombia and Peru. *Latin American Politics and Society, 51*(1), 57–85. https://doi.org/10.1111/j.1548-2456.2009.00040.x

Baeza-Freer, J., & Escudero-Illanes, M. C. (2024). Sendero dependencia en presencia de militares chilenos en catástrofes. *Revista Científica General José María Córdova, 22*(48), 1103–1124. https://doi.org/10.21830/19006586.1385

Barbosa Castillo, G., & Ciro Gómez, A. R. (Eds.). (2020). *Garantía de no repetición. Una contribución a la justicia transicional.* Universidad Externado de Colombia.

Barrachina-Lisón, C., & Benítez-Manaut, R. (2024). México. Relaciones civiles-militares y desastres naturales. *Revista Científica General José María Córdova, 22*(48), 1085–1102. https://doi.org/10.21830/19006586.1386

Bartolomé, M. C. (2018). La Seguridad Internacional contemporánea: contenidos temáticos, agenda y efectos de su ampliación. *Relaciones Internacionales, 55*, 123–145.

Besio-Echeverría, F. A. (2024). Utilización de las Fuerzas Armadas uruguayas en situaciones de emergencia y desastres. *Revista Científica General José María Córdova, 22*(48), 1067–1084. https://doi.org/10.21830/19006586.1387

Bobea, L. (2024). Las Fuerzas Armadas dominicanas y su participación en la gestión de emergencias. *Revista Científica General José María Córdova, 22*(48), 1045–1065. https://doi.org/10.21830/19006586.1388

Born, H., Caparini, M., & Fluri, P. (Eds.). (2000). *Security Sector Reform and Democracy in Transitional Societies.* Geneva Centre for the Democratic Control of Armed Forces. https://tinyurl.com/4dbn3449

Bucur-Marcu, H., Fluri, P., & Tagarev, T. (Eds.). (2013). *Gestión de la defensa. Una Introducción.* Centro de Ginebra para el Control Democrático de Fuerzas Armadas.

Bueno, A., & Martínez, R. (2024). La militarización de las emergencias civiles en Guatemala. *Revista Científica General José María Córdova, 22*(48), 1023–1044. https://doi.org/10.21830/19006586.1390

Cabrera-Ortiz, F. (2021). Propuesta para el planeamiento estratégico de la seguridad nacional desde una perspectiva multidimensional. *Revista Científica General José María Córdova, 19*(33), 5–28. https://doi.org/10.21830/19006586.747

Calduch, R. (2020). ¿Cultura de defensa o cultura de seguridad nacional? Un nuevo debate doctrinal en España. *Revista de Ensayos Militares, 6*(2), 71–95.

Cano Cuevas, D. F. (2024). La seguridad y defensa nacional y su relación con el derecho operacional. In *Abordando los conflictos contemporáneos: construcción de paz, posconflicto y el paradigma de la seguridad* (pp. 89–128). Escuela Superior de Guerra. https://doi.org/10.25062/9786287602786.04

Cardenas, S., & Root, R. K. (2022). *Human Rights in Latin America: A Politics of Transformation.* University of Pennsylvania Press.

Cepeda-Jaramillo, A., Leetoy, S., & Zavala-Scherer, D. (2021). Reactivación mítica, performance, y agencia cultural: El caso de las Madres de los Falsos Positivos de Colombia. *Revista de Estudios Colombianos, 57.* https://doi.org/10.53556/rec.v57i0.155

Chavez Cresta, J. L. (2021). *Fuerzas armadas y la gestión del riesgo de desastres en América Latina y el Caribe.* CEEEP.

Cremades, A. (2021). Sobre las instituciones estatales de seguridad en América Latina: similitudes y particularidades frente al caso español. *Methaodos Revista de Ciencias Sociales, 9*(1), 45–59. https://doi.org/10.17502/mrcs.v9i1.447

Dammert, L. (2019). ¿Reformar sin gobernar? Desafíos institucionales de las policías en América Latina. *Revista IUS, 13*(44), 89–120. https://doi.org/10.35487/rius.v13i44.2019.444

Day, C., & Khisa, M. (2022). Rethinking Civil-Military Relations in Africa. In *Rethinking Civil-Military Relations in Africa* (pp. 1–38). Lynne Rienner Publishers. https://doi.org/10.1515/9781955055468-003

DCAF. (2015a). *La Sociedad Civil. Funciones y responsabilidades en la buena gobernanza del sector de seguridad.* DCAF. https://tinyurl.com/4u3huax2

DCAF. (2015b). *Reforma del Sector de Seguridad. Aplicación de los principios de buena gobernanza al sector de seguridad.* DCAF. https://tinyurl.com/mc3v7bcf

De Vicente Encarnación, J. (2024). *Reform of Mexico's National Guard: Towards Total Militarization.* Wilson Center. https://tinyurl.com/5n83u8sx

Diamint, R. (2015). A New Militarism in Latin America. *Journal of Democracy, 26*(4), 155–168. https://doi.org/10.1353/jod.2015.0066

Droz-Vincent, P. (2014). Prospects for "Democratic Control of the Armed Forces"? *Armed Forces & Society, 40*(4), 696–723. https://doi.org/10.1177/0095327X12468881

Expósito, E. (2024). El rol de los militares en la gestión de las emergencias. *Revista Científica General José María Córdova, 22*(48), 1003–1021. https://doi.org/10.21830/19006586.1391

Feaver, P. (1998). Crisis as Shirking: An Agency Theory Explanation of the Souring of American Civil-Military Relations. *Armed Forces & Society, 24*(3), 407–434. https://doi.org/10.1177/0095327X9802400305

Feaver, P. (2003). *Armed servants : agency, oversight, and civil-military relations.* Harvard University Press.

Fernandez-Osorio, A. E. (2021). *Espartanos y Atenienses: Un estudio contemporáneo de la identidad militar en Colombia.* Sello Editorial ESMIC. https://doi.org/10.21830/9789585350625

Finer, S. E. (1969). The Military in the Politics of Today. In *Comparative Government* (pp. 208–216). Macmillan Education UK. https://doi.org/10.1007/978-1-349-15318-3_24

Galindo García, F. L., & Aristizábal Murillo, J. C. (2021). The United Nations and 21st century security challenges in Colombia. *Revista Científica General José María Córdova, 19*(36), 929–940. https://doi.org/10.21830/19006586.875

Germani, G., & Silvert, K. (1961). Politics, Social Structure and Military Intervention in Latin America. *European Journal of Sociology, 2*(1), 62–81. https://doi.org/10.1017/S000397560000028X

Gobierno de Chile. (2019). *Política nacional de ciberseguridad*. Comité Interinstitucional sobre Ciberseguridad. https://tinyurl.com/24vuw3vm
Godnick, W. H. (2023). *Herramientas de Prospectiva de Defensa y Seguridad Nacional de Estados Unidos y Australia y su aplicabilidad al contexto de América Latina*. William J. Perry Center for Hemispheric Defense Studies.
Gorbanova, M., & Wawro, L. (2011). *The Transparency of National Defence Budgets*. Transparency International UK. https://tinyurl.com/y79hw53r
Guzmán Mesa, V. (2016). *The Colombian emergency response platform (PCRE): design and testing of a sdi-based system to facilitate disaster management in Colombia* [Universidad de los Andes]. https://tinyurl.com/t4s8yc6s
Hamilton, M., & Bordón, R. P. (2023). ¿La crónica de una militarización anunciada? El rol militar y las relaciones cívico-militares en los tiempos de pandemia. *Perspectivas Revista de Ciencias Sociales*, *8*(15), 1–26. https://tinyurl.com/2w3p5sk2
Holshek, C., & Coning, C. de. (2024). *Civil-Military Coordination in Peace Operations*. Peace Operations Training Institute. https://tinyurl.com/2y3mhjup
Huntington, S. P. (1957). *The soldier and the State: the theory and politics of civil-military relations*. Harvard University Press.
Inter-American Defense Board. (2024a). *Confidence-Building Measures and Military Security*. Line of Action. https://tinyurl.com/2ddmmzt4
Inter-American Defense Board. (2024b). *Disaster Cooperation*. Line of Action. https://tinyurl.com/yc69srw6
Jaimes Pinillos, J. H. (2018). Introducción a la sociología militar. In H. Martínez Espinel (Ed.), *Perspectiva histórica de la sociología militar* (pp. 27–43). Sello Editorial ESMIC. https://doi.org/10.21830/9789585962712
Janot, M. da G., & de Araujo de Assis, J. (2025). Experimenting with Drones: Remotely Piloted Aircraft and Security Practices in Brazil. *Global Studies Quarterly*, *5*(2). https://doi.org/10.1093/isagsq/ksaf030
Janot, M. da G., Succi-Junior, D. P., & Soares, S. A. (2024). The armed forces role in emergencies in Brazil. *Revista Científica General José María Córdova*, *22*(48), 983–1001. https://doi.org/10.21830/19006586.1392
Janowitz, M. (1974). *The professional soldier: a social and political portrait*. The Free Press.
Jaskoski, M. (2012). Civilian Control of the Armed Forces in Democratic Latin America. *Armed Forces & Society*, *38*(1), 70–91. https://doi.org/10.1177/0095327X11398449
Jenne, N., & Martínez, R. (2022). Domestic military missions in Latin America: Civil-military relations and the perpetuation of democratic deficits. *European Journal of International Security*, *7*(1), 58–83. https://doi.org/10.1017/eis.2021.25
Jiménez Espinosa, P. V., & Espinosa Gallardo, J. O. (2020). Las fuerzas armadas en apoyo a desastres naturales. *Revista de La Academia Del Guerra Del Ejército Ecuatoriano*, *13*(1), 10. https://doi.org/10.24133/age.n13.2020.09
Jorquera Escobar, R. (2024). Tecnología e inteligencia artificial, desafíos para la defensa. *Centro de Estudios Estratégicos y Aeroespaciales*, *21*, 1–18. https://tinyurl.com/56s8n3sj
Krupski, K. F. (2023). ¿Quién es el jefe? Definir las relaciones cívico-militares en el siglo XXI. *Military Review*, *3*, 10–19. https://tinyurl.com/muy3j2uz
Longpre, S., Storm, M., & Shah, R. (2022). Lethal autonomous weapons systems & artificial intelligence: Trends, challenges, and policies. *MIT Science Policy Review*, *3*, 47–56.
Luque-Juárez, J. M. (2024). La seguridad humana y su implementación en la operación de las Fuerzas Armadas. *Revista Científica General José María Córdova*, *22*(45), 237–258. https://doi.org/10.21830/19006586.1237
Marczak, J., Engelke, P., Bohl, D., & Saldarriaga Jiménez, A. (2016). *América Latina y el Caribe 2030: Escenarios futuros*. Banco Interamericano de Desarrollo. https://tinyurl.com/4m4u4zdf
McSherry, J. P. (1995). Military political power and guardian structures in Latin America. *Journal of Third World Studies*, *12*(1), 80–119.

Micha, L., & Farías, P. (2021). *Los riesgos de las armas autónomas*. Asociación para Políticas Públicas. https://tinyurl.com/2n77h4uk

Ministerio de Defensa Nacional. (2013). *Primer libro blanco de la defensa nacional de la república del Paraguay*. Ministerio de Defensa Nacional. https://tinyurl.com/3pnpm7yn

Mockaitis, T. R. (2004). *Civil-military cooperation in peace operations: The case of Kosovo*. US Army War College.

Molina-Gabriel, R. A. (2024). La Fuerza Armada de El Salvador y su tradicional versatilidad en emergencias. *Revista Científica General José María Córdova*, *22*(48), 963–982. https://doi.org/10.21830/19006586.1393

Moncayo Gallegos, P. (2018). La política de seguridad y defensa. *Revista de La Academia Del Guerra Del Ejército Ecuatoriano*, *11*(1), 8–18. https://tinyurl.com/wzd2r438

Monroy Hernández, M. C., Ortega Gómez, A. F., Díaz Vallenoti, G. A., & Mesa Payares, J. (2013). Una aproximación a los estudios de seguridad ciudadana en Caracas. *Revista Científica General José María Córdova*, *11*(11), 155. https://doi.org/10.21830/19006586.209

Morlino, L. (2014). *La calidad de las democracias en America Latina*. Instituto Internacional para la Democracia y la Asistencia Electoral. https://tinyurl.com/2s387kpv

Moskos, C. C. (1986). Institutional/Occupational Trends in Armed Forces: An Update. *Armed Forces & Society*, *12*(3), 377–382. https://doi.org/10.1177/0095327X8601200303

Muggah, R. (2017). The Rise of Citizen Security in Latin America and the Caribbean. *Revue Internationale de Politique de Développement*, *9*, 291–322. https://doi.org/10.4000/poldev.2377

Nielsen, S. C., & Liebert, H. (2021). The Continuing Relevance of Morris Janowitz's The Professional Soldier for the Education of Officers. *Armed Forces & Society*, *47*(4), 732–749. https://doi.org/10.1177/0095327X20960480

Novak-Talavera, F. (2024). El rol de las Fuerzas Armadas del Perú en caso de desastres naturales. *Revista Científica General José María Córdova*, *22*(48), 943–961. https://doi.org/10.21830/19006586.1394

Orozco, G. (2005). El concepto de la seguridad en la Teoría de las Relaciones Internacionales. *CIDOB d'Afers Internacionals*, *72*, 161–180.

Pacheco Gaitán, G. (2016). Los Libros Blancos: ¿Un paso para consolidar medidas de confianza? *Security and Defense Studies Review*, *2*(1), 101–115. https://tinyurl.com/dp9je6pt

Perito, R. M. (2019). Security sector reform. Applying the principles of good governance to the security sector. *SSR Backgrounder*, *1*, 1–8. https://doi.org/10.4337/9781788971645.00017

Pion-Berlin, D. (2016). *Military Missions in Democratic Latin America*. Palgrave Macmillan.

Presidencia de la República Dominicana. (2023). *Presidente Luis Abinader dice el Libro Blanco de la Defensa de RD es instrumento vivo para enriquecer con la sociedad*. Seguridad. https://tinyurl.com/vr6xt9c5

Quintero Cordero, S. P. (2020). Seguridad ciudadana y participación de las comunidades en América Latina. *Revista Científica General José María Córdova*, *18*(29), 5–24. https://doi.org/10.21830/19006586.561

Remmer, K. (2025). *Military Rule in Latin America*. Routledge. https://doi.org/10.4324/9781003440727

Resdal. (2018). *Ecuador. Libro Blanco de la Defensa Nacional - Las Fuerzas Armadas*. https://tinyurl.com/4h9ax7ss

Robledo Hoecker, M. (2022). Militarización, emergencia del militarismo civil y erosión democrática en América Latina. *Documentos de Trabajo*, *74*, 1–31.

Rodriguez, S. M. (2025). Behind the trigger: democracy, neoliberalism, and civilian militarism in Latin America. *Third World Quarterly*, *46*(1), 1–19. https://doi.org/10.1080/01436597.2024.2433698

Romero Herrera, J. (2007). Relaciones Civiles-Militares en América latina, el control civil sobre los militares y la ampliación de las actividades profesionales de los militares. *Estudios En Seguridad y Defensa*, *2*(3), 6. https://doi.org/10.25062/1900-8325.150

Ross, J. N. (2004). *The Changing Role of the Military in Latin America*. The Canadian Foundation for the Americas.

Rouquié, A. (1987). *The military and the State in Latin America*. University of California Press.
Ruiz Campillo, X., & Montenegro Escánez, F. (2019). *Seguridad y Defensa en Latinoamérica: de los Libros Blancos de la Defensa a la Cooperación Regional*. Fundación Alternativas y Ministerio de Defensa. https://tinyurl.com/bdzafeut
Sánchez, W. A. (2014). COHA Report: Drones in Latin America. *COHA Policy Memo*, *4*, 1–10.
Sandhu, I. (2016). Finer's Conditions of Military Intervention: The Case of the 1999 Coup in Pakistan. *Carleton Review of International Affairs*, *3*. https://doi.org/10.22215/cria.v3i0.121
Schiff, R. L. (1995). Civil-Military Relations Reconsidered: A Theory of Concordance. *Armed Forces & Society*, *22*(1), 7–24. https://doi.org/10.1177/0095327X9502200101
Schiff, R. L. (2008). *The Military and Domestic Politics: A Concordance Theory of Civil-Military Relations*. Routledge. https://doi.org/10.4324/9780203892305
Schmidt, T. (2024). Frágil y brutal. Un pronóstico sobre relaciones cívico-militares para el 2024 que podría evitarse. *Military Review*, *3*, 2–5.
Segal, D. R. (1986). Measuring The Institutional/Occupational Change Thesis. *Armed Forces & Society*, *12*(3), 351–375. https://doi.org/10.1177/0095327X8601200302
Silva, P. (2002). Democratisation and state-civil society relations in Chile, 1983–2000: from effervescence to deactivation. *Iberoamericana. Nordic Journal of Latin American and Caribbean Studies*, *XXXII*(2), 73–96.
Skaar, E., & Malca, C. G. (2014). Latin American civil-military relations in a historical perspective: A literature review. *Working Paper - Chr. Michelsen Institute*, *6*, 1–30.
Soprano, G. (2010). Los militares como grupo social y su inscripción en el estado y la sociedad argentina. Batallas intelectuales y políticas por la construcción de un objeto de estudio en las ciencias sociales. *Revista Digital Universitaria Del Colegio Militar de La Nación*, *22*, 1–28. https://tinyurl.com/y55xtvar
Sotomayor Velásquez, A. (2006). Diagnóstico de las relaciones cívico-militares en América Latina: Avances y retrocesos en materia de política de defensa. *Documentos de Trabajo-CIDE*, *142*, 1–37. https://tinyurl.com/4r8d454s
Stepan, A. C. (2001). *Arguing comparative politics*. Oxford University Press.
Stepan, A. C. (2020). *Rethinking Military Politics: Brazil and the Southern Cone*. Princeton University Press.
Stone, H. (2019). *Corruption and Plan Colombia: The Missing Link*. Transparency International UK. https://tinyurl.com/yzuyx6hy
Sung, H.-E., Capellan, J., & Barthuly, B. (2022). Trust in the police and the militarisation of law enforcement in Latin America. *Policing and Society*, *32*(3), 311–340. https://doi.org/10.1080/10439463.2022.2037551
Taylor, J. K., Clerkin, R. M., Ngaruiya, K. M., & Velez, A.-L. K. (2015). An Exploratory Study of Public Service Motivation and the Institutional–Occupational Model of the Military. *Armed Forces & Society*, *41*(1), 142–162. https://doi.org/10.1177/0095327X13489119
Telleria, L. (2024). Estados Unidos y el proceso de militarización de América Latina. *Reoriente: Estudos Sobre Marxismo, Dependência e Sistemas-Mundo*, *3*(2), 119–143. https://doi.org/10.54833/issn2764-104X.v3i2p119-143
Tirado, A. (2021). *El lawfare: Golpes de Estado en nombre de la ley*. Ediciones Akal.
Travis, D. S. (2017). Saving Samuel Huntington and the Need for Pragmatic Civil–Military Relations. *Armed Forces & Society*, *43*(3), 395–414. https://doi.org/10.1177/0095327X16667287
Travis, D. S. (2020). Decoding Morris Janowitz: Limited War and Pragmatic Doctrine. *Armed Forces & Society*, *46*(1), 68–91. https://doi.org/10.1177/0095327X18760272
Ugarte, J. M. (2024). El rol de las Fuerzas Armadas argentinas en emergencias y el apoyo a la comunidad. *Revista Científica General José María Córdova*, *22*(48), 925–942. https://doi.org/10.21830/19006586.1395
Uluçakar, M., & Çağlar, A. (2017). An Analysis of Two Different Models of Civil-Military Relations: The Case of Turkey. *Uluslararası İlişkiler Dergisi*, *14*(55), 41–57. https://doi.org/10.33458/uidergisi.513504

United Nations. (2023). *Security sector reform.* Building Rule of Law & Security Institutions. https://tinyurl.com/yfydcw3n

Velázquez, A. C. S. (2008). Los civiles y militares en América Latina: avances y retrocesos en materia de control civil. *Revista Mexicana de Política Exterior*, *83*, 41–84.

Victoriano Serrano, F. (2010). Estado, golpes de Estado y militarización en América Latina: una reflexión histórico política. *Argumentos*, *23*(64), 175–193.

Villalba-Garcia, L. F. (2022). Economía de defensa y política pública en Latinoamérica y el Caribe. In *Economía de defensa, políticas públicas y resolución de conflictos* (pp. 77–89). Sello Editorial ESDEG. https://doi.org/10.25062/9786287602243.07

Huntington, S. P. (1991). The third wave: Democratization in the late twentieth century. University of Oklahoma Press.

Buzan, B., Wæver, O., & de Wilde, J. (1998). Security: A new framework for analysis. Lynne Rienner Publishers.

Andres Eduardo Fernandez-Osorio Ph.D. in Law and Political Science, University of Barcelona, Spain. MA in Economics, State and Society: Politics and Security, University College London, United Kingdom. MA in International Relations, Higher School of Economics, Russian Federation. MA in Defense Studies, King's College London, United Kingdom. BA in International Relations and Political Studies, Universidad Militar Nueva Granada, Colombia. B.Sc. in Military Sciences, Escuela Militar de Cadetes General José María Córdova, Colombia. Adjunct Professor and researcher at the School of Humanities, Universidad Gabriela Mistral, Santiago, Chile. https://orcid.org/0000-0003-0643-0258—Contact: andres.fernandez@academico.ugm.cl.

Luisa Fernanda Villalba-Garcia Ph.D. (c) in Strategic Studies, Security and Defense and MA in Strategy and Geopolitics, Escuela Superior de Guerra "General Rafael Reyes Prieto," Colombia. BA in International Relations and Political Studies, Universidad Militar Nueva Granada, Colombia. Professor and researcher at the Escuela Militar de Cadetes "General José María Córdova", Colombia. https://orcid.org/0000-0003-3169-9312—Contact: luisa.villalba@esmic.edu.co

Christian Acevedo-Navas Ph.D. in Social Sciences, Universidad del Norte, Colombia. B.Sc. in Naval Sciences, Escuela Naval de Cadetes "Almirante Padilla". Professor and researcher on the PhD program in Strategic Studies, Security, and Defense, Escuela Superior de Guerra "General Rafael Reyes Prieto." https://orcid.org/0000-0003-4880-3024—Contact: christian.acevedo@esdeg.edu.co.

MIX
Papier aus verantwortungsvollen Quellen
Paper from responsible sources
FSC® C105338

If you have any concerns about our products,
you can contact us on
ProductSafety@springernature.com

In case Publisher is established outside the EU,
the EU authorized representative is:
Springer Nature Customer Service Center GmbH
Europaplatz 3, 69115 Heidelberg, Germany

Printed by Libri Plureos GmbH
in Hamburg, Germany